CONTRIBUTIONS TO THE GENETICS
OF PISUM

Reprint from: Genetica XII, 1930

Contributions to the genetics
of
PISUM

BY

Dr. H. DE HAAN

WITH 11 FIGURES AND 2 COLOURED PLATES

'S-GRAVENHAGE
MARTINUS NIJHOFF
1931

ISBN 978-94-015-2070-6   ISBN 978-94-015-3277-8 (eBook)
DOI 10.1007/978-94-015-3277-8

CONTRIBUTIONS TO THE GENETICS OF PISUM

by

H. DE HAAN

(With 11 figures and 2 coloured plates)

TABLE OF CONTENTS

INTRODUCTION

A number of investigators have chosen the pea as experimental plant for genetical researches and in 1925 WELLENSIEK (91, p. 343) gave a summary of the state of the research on heredity. To subsequent researches this monograph was a great help.

In spite of the many publications which have appeared of late years on the heredity in the pea, but a few of the hereditary differences have been analysed, while the research on the localization of the factors is still in its initial stage. So the pea of which so many forms occur, for the time being presents plenty of material to start a research.

I have set myself the task of making the completest possible analysis of the flower colour differences and of the differences in length of stem. As regards the flower colour I build on the researches of H. TEDIN, who was kind enough to spare me his forms for my research. In a way it is a happy coincidence that the number of flower colours of the pea is rather small, which renders the research surveyable. Of the known forms I now possess a fairly complete collection, pains being constantly taken to supplement the collection. The research on the heredity of the length of stem is still in its initial stage. So vast a number of forms exist that I have more material at my disposal than can be worked up.

The researches were made in the Genetic Institute of the State University of Groningen.

CHAPTER I

FLOWER COLOUR FACTORS

§ 1. Introduction

MENDEL found that the difference between purple- and white flowering peas is caused by a single factor (59, 1866, p. 14). Not until after 1900, however, the flower colour of the pea has been subjected to an extensive genetic examination. To the colours studied rose, light purple and violet were added. In the various publications, however, not the same symbols are used for the hereditary differences. For this reason the TEDINS and WELLENSIEK (84, 1925, p. 533) agreed to make them fit in with each other and to indicate the three determined factors by A, A_r and B. The action of these factors was described by them as follows:

A : fundamental factor for flower colour, giving light purple by itself.

A_r: together with A, without B, gives pink (rose).

B : together with A gives violet, together with A and A_r, purple.

Here A is the factor for the difference between purple and white originally determined by MENDEL.

The formulas for the flower colour are as follows:

$AA\ A_r\ A_r\ BB$ purple (in older publications sometimes called 'rotviolett', 'violettrot' or 'rot'),

$AA\ a_r\ a_r\ BB$ violet,

$AA\ A_r\ A_r\ bb$ rose (sometimes called '*Lathyrus* flowered' or 'salmon pink'),

$AA\ a_r\ a_r\ bb$ light purple,

$aa\ A_r\ A_r\ BB$ white (for the rest all aa forms are white, no matter whether A_r and B occur or not).

The cultivated white flowered forms, however, are almost without exception of the constitution $aa\ A_r\ A_r\ BB$.

Pictures of the variously coloured flowers were given by TEDIN (81, 1920, p. 98) and represented by TSCHERMAK on plates 1 to 4 (88, 1913, p. 573).

In the old herbals purple and white (OBEL, 64, 1581, p. 77) are mentioned while I found rose mentioned for the first time in MILLER (60, 1786, tome 6, p. 16). Light purple was received in 1898 by TEDIN as 'Ljusröd blommig' while violet arose in his crosses about 1911 (81, 1920, p. 69).

§ 2. *Apple blossom*

Among the cultivated forms of the grey peas varieties occur with a more or less flesh coloured flower, the standard of which is also brightly coloured, so that this type gives the impression of being self coloured. Besides in the flower colour, therefore, these forms also differ in their being unicoloured from the other forms mentioned in § 1, which are designated bicoloured.

The growers call these unicoloured forms apple blossom or 'bleek-bloeier' ('pale bloomer'). Below, however, I shall only use the name apple blossom (pl. I, fig. 2).

Apple blossom appeared to be a constant form. The progeny of the type cultivated from trade-seed were all pure.

In order to trace the hereditary difference between apple blossom and the other forms, different crossings were made. It appeared from these that the F_1's of apple blossom and purple, white, rose or violet were invariably purple.

Crossing purple with apple blossom gave the following results:

TABLE 1. Apple blossom × purple

year	No.	purple	apple blossom	total
1929	303	84	32	116
„	93	179	64	243
total		263	96	359

theor. 3 : 1   269.25 : 89.75
     m   8.11
     D/m   0.771

From these observations it may be concluded that purple and apple blossom differ in 1 factor. Since purple is of a constitution AA A_r A_r BB, it may be that the difference is due to one of these factors. As, however, all the combinations in which one of these factors does not occur, are known, a new hereditary factor may be concluded to, which will be designated A_p. A s purple dominates, it may be accepted that purple possesses A_p, apple blossom a_p. Besides we may conclude from the difference in 1 factor that for the rest apple blossom has the same factors as purple and consequently possesses the formula AA A_r A_r BB a_p a_p.

In order to test the correctness of this apple blossom was crossed with white (aa A_r A_r BB). The data of the F_2 are given in the subjoined table.

TABLE 2. White × apple blossom

year	No.	purple	apple blossom	white	total
1929	96	197	69	88	354

theor. 9 : 3 : 4   199.1 : 66.4 : 88.5
     m   9.28   7.31   8.15
     D/m   0.226   0.356   0.061

Since purple and white differ in the factor A, purple and apple in the factor A_p, it may be expected that between apple and white there should exist a bifactorial difference. As appears from the table this is, indeed, the case, while from the ratio 9 : 3 : 4 an independent transmission of A and A_p may be inferred. The double recessive $aa\ a_p\ a_p$ forms are white, just as is the case with other forms in which the ground factor for colour does not occur.

Apple blossom was also crossed with rose. Here too the two differ in one factor from purple, so that a dihybrid segregation may be expected. That the observations are in accordance with expectation appears from the subjoined table.

TABLE 3.   Rose × apple blossom

year	No.	purple	rose	apple blossom	apple rose	total
1929	94	234	73	78	23	408
theor.						
9 : 3 : 3 : 1		229.5 :	76.5 :	76.5 :	25.5	
	m	10.12	7.92	7.86	4.90	
	D/m	0.444	0.442	0.191	0.510	

From the ratio 9 : 3 : 3 : 1 the independent transmission of B and A_p may be inferred. In this F_2 a new form arose that is the double recessive $bb\ a_p\ a_p$, which is self coloured just as apple blossom, but differs from it in the flower colour being rose apple blossom. These 'apple rose' plants are to be subdivided into deep apple rose (pl. I, fig. 1) and dilute apple rose. Of both forms 2 plants were further cultivated in 1930 and it was ascertained that they were pure for colour. Though further examination will teach what the hereditary difference in deep and dilute apple rose is based upon, it may be established that it cannot be explained from the factors known at present. The same may be observed with regard to the rose plants from the same F_2, as these are to be subdivided into dilute rose and deep rose (called red). In how far it is the same factor which causes this subdivision is still being investigated.

Likewise the crossing violet ($a_r\ a_r\ BB$) by apple blossom was made. As I only possessed violet F_2 plants from the cross light purple and white, the possibility existed that these should be heterozygous for B. From table 4 follows a bifactorial segregation, while from the 9 : 3 : 3 : 1

TABLE 4. Violet × apple blossom

year	No.	purple	violet	apple	apple violet	total
1930	90	68	21	23	6	118

theor. 9 : 3 : 3 : 1	66.3 :	22.1 :	22.1 :	7.4
m	5.45	4.26	4.22	2.64
D/m	0.312	0.258	0.213	0.530

ratio appears, that A_r and A_p are transmitted independently. In the F_2 a new form arose, viz. the double recessive $a_r\, a_r\, a_p\, a_p$, which was called apple violet and possessed a colour preserving the mean between violet and apple blossom. From the segregation it might be concluded that the F_1 was homozygous for B. This was not the case with the next, where as a result of heterozygotism for the factor B, rose, apple rose and light purple also occurred. From the observations given in table 5 a trifactorial segregation may be inferred. The form

TABLE 5. Violet × apple blossom

year	No.	pur-ple	viol.	apple	rose	apple viol.	apple rose	light purple	?	total
1930	91	57	14	18	18	5	7	3	—	122

theor. 27 : 9 : 9 : 9 : 3 : 3 : 3 : 1								
	51.5 :	17.2 :	17.2 :	17.2 :	5.7 :	5.7 :	5.7 :	1.4
m	5.24	3.89	3.82	3.82	2.34	2.32	2.36	1.37
D/m	1.049	0.823	0.209	0.209	0.299	0.560	1.144	1.022

which is recessive for a_r, b and a_p has not occurred. From what has been communicated, however, about the interaction of the A_p factor and the other factors it may be foretold that this expected form will preserve the mean between apple blossom and light purple.

From the agreement of the results of crossing with the theoretical expectation it may be concluded, that purple, rose, white and violet possess the factor A_p homozygously, while it may be inferred that this is likewise the case with light purple. Apple blossom, apple rose and apple violet are recessive for the A_p factor.

§ 3. *Pinkish white*

Through the kind intermediary of Dr. R. J. MANSHOLT (Westpolder, Groningen) I received seed of a form possessing a different genotypical constitution from all the above-mentioned forms. About the origin of the seed Dr. MANSHOLT informed me that in 1922 he sowed out a sample of trade-seed of a population of short grey peas from North-Holland. The seed appeared to be a mixture, as in the plot of ½ are (50 square meters) beside the common short plants there occurred 5 tall plants and also 5 short plants with white flowers. These latter were harvested separately. They appeared to be plants deviating from the rule that coloured flowers and a coloured seedcoat go together. The flowers were white, the seeds coloured. The remaining purple flowering grey peas appeared to be constant in the subsequent years, so that the chances are that the aberrant plants were not due to segregation but to admixture in the trade-seed.

However, the plants from the peas received appeared not to be pure white flowering, the flowers showing a red colour after treatment with HCl. Pure white (*aa*) flowers on the other hand do not produce colour on treatment with an acid. As to the flower colour, however, the former cannot as a rule be directly distinguished from white, sometimes when a bud they show a faint pink colour at the margins of the wings. In connection with this the new flower colour was called pinkish white.

The progeny in the years 1925, 1926, etc. consisted of plants of the same type: pinkish white flowers and coloured seeds, so that we have to do here with a new constant form. In order to trace the genotypical constitution various crossings were made.

The F_1 of the cross purple × pinkish white proved to be purple, this being also the case when rose, white, violet, light purple, apple blossom and apple rose were used for crossing. From such a behaviour it may be inferred that pinkish white possesses all known factors, while the well-known flower colour types will possess the very factor or factors, in which pinkish white differs from purple. This supposition is confirmed by the crossing-results obtained.

From the cross pinkish white × purple it appears that the difference from purple is monofactorial. The data in table 6 tally fairly well with the theoretical ratio 3 : 1. As, however, of the known factors A, A_r, B and A_p all forms which differ from purple in one of these

TABLE 6. Pinkish white ~ purple

year	No.	purple	pinkish white	total
1926	5	103	32	135
,,	6	87	30	117
,,	7	107	37	144
,,	13	275	90	365
,,	14	231	84	315
,,	17	78	21	99
,,	18	63	20	83
,,	19	61	16	77
,,	20	68	25	93
,,	21	68	21	89
,,	22	52	15	67
,,	23	63	21	84
total		1256	412	1668

theor. 3 : 1   1251.0   :   417.0

m   17.72   D/m   0.282

factors are known, it may be assumed that an other factor determines the difference. This factor was called A_m and to A_m the action was attributed of inhibiting anthocyanin formation in a recessive condition. Consequently pinkish white has all other factors in common with purple and its formula must be $AA\ A_r\ A_r\ BB\ A_p\ A_p\ a_m\ a_m$. The factors A, A_r, B and A_p cannot manifest themselves in the flower colour, because the factor A_m does not occur. From this it appears that A_m exercises a very great influence on the colour formation and shows a great correspondence with a ground factor. If this supposition is correct, it may be expected just as in crossings with white (aa), that all $a_m\ a_m$ forms are pinkish white, independent of the further colour factors.

The above suppositions are confirmed by the following crosses.

The cross pinkish white with white (aa) shows a bifactorial difference, as may be expected, since both forms differ from purple in one factor and moreover it may be inferred from the 9 : 3 : 4 ratio in the subjoined table 7 that A and A_m are transmitted independently.

The crossing of pinkish white and rose yielded results which fully agree with the expectation that they differ in the factors B and A_m.

The data are given in the subjoined table 8. From the agreement with

TABLE 7.   Pinkish white ∼ white

year	No.	purple	pinkish white	white	total
1927	8.1	139	44	68	251
,,	8.2	112	37	44	193
,,	8.3	89	26	40	155
,,	9.1	114	40	49	203
,,	9.2	129	41	61	231
total		583	188	262	1033
theor. 9 : 3 : 4		581.1 :	193.7 :	258.2	
m		15.97	12.59	13.88	
D/m		0.119	0.453	0.274	

the 9 : 3 : 4 segregation independent transmission of B and A_m may be concluded to. In this F_2 a new form did not occur, because the double recessive $bb\ a_m\ a_m$ is not to be distinguished from pinkish white.

TABLE 8.   Rose × pinkish white

year	No.	purple	rose	pinkish white	total
1929	98	208	69	81	358
1930	135	162	51	67	280
total		370	120	148	638
theor. 9 : 3 : 4		358.9 :	119.6 :	159.5	
m		12.72	9.85	11.07	
D/m		0.873	0.041	1.039	

In the cross pinkish white × apple blossom both differing from purple in one factor, a bifactorial difference may be expected. Here too the new double recessive form cannot be distinguished from pinkish white. From the 9 : 3 : 4 ratio occurring in the subjoined table the independent transmission of A_p and A_m may be inferred.

In the crosses of pinkish white discussed above new flower colours have not originated. This confirms the supposition that A_m in a

TABLE 9. Apple blossom × pinkish white

year	No.	purple	apple blossom	pinkish white	total
1929	305	129	44	61	234
theor. 9 : 3 : 4		131.6 :	43.9 :	58.5	
m		7.51	5.97	6.58	
D/m		0.346	0.017	0.380	

recessive condition has an inhibitory effect on the formation of antho-cyanin. Besides we may conclude from the above that all known flower colour types possess constitution $A_m A_m$ and that among the pinkish white forms with the factors $a_m a_m$ different genotypes may occur in the same way as in the case of forms in which the ground factor for colour A does not occur.

§ 4. The flower colour formulas

From the above it has appeared that the flower colour formulas have to be completed with 2 factors A_p and A_m. Besides it follows that the description of the factors by the TEDINS and WELLENSIEK (84, 1925, p. 533) should be modified in connection with this. The effect of the ground factor A was inferred by them from the colour of the $A A a_r a_r bb$ form, in which A occurs by itself. From this form it was concluded that A by itself gives light purple.

To this I must raise objections. Originally it was inferred from the rose $A A bb$ form known at the time that A by itself produce rose. According as more factors become known, a different action will be attributed to A. This objection does not exist, when we start with the recessive factors and we say in the case of rose that forms in which B does not occur are rose, or in the first mentioned case that forms in which A_r and B do not occur, are light purple. In doing so we take into account the possibility that besides $A A$ still other factors are the common property of all flower colours.

From the fact that all aa forms are pure white, nor produce any colour on addition of an acid, it may be concluded that A is a ground factor that must be present in order to enable other colour factors to manifest themselves. A by itself does not produce colour. And if it might appear that the $A A a_r a_r bb a_p a_p$ form to be expected in the cross light purple with apple blossom shows colour, this colour must not be attributed to the factor A, but it may only be concluded that forms in which A_r, B and A_p do not occur are apple light purple, if we give this name to $AA\ a_r\ a_r\ bb\ a_p\ a_p$. Moreover it may be foretold that this apple light purple colour arises because there are more colour factors. We shall not be able to place these factors in the formulas until they are made open to investigation by factor mutation.

The formulas of the known forms are:

purple	$AA\ A_r\ A_r\ BB\ A_p\ A_p\ A_m\ A_m$
violet	$AA\ a_r\ a_r\ BB\ A_p\ A_p\ A_m\ A_m$ (pl. I, fig. 3)
rose	$AA\ A_r\ A_r\ bb\ A_p\ A_p\ A_m\ A_m$ (pl. I, fig. 4)
light purple	$AA\ a_r\ a_r\ bb\ A_p\ A_p\ A_m\ A_m$
apple blossom	$AA\ A_r\ A_r\ BB\ a_p\ a_p\ A_m\ A_m$ (pl. I, fig. 2)
apple rose	$AA\ A_r\ A_r\ bb\ a_p\ a_p\ A_m\ A_m$ (pl. I, fig. 1)
apple violet	$AA\ a_r\ a_r\ BB\ a_p\ a_p\ A_m\ A_m$
pinkish white	$AA\ A_r\ A_r\ BB\ A_p\ A_p\ a_m\ a_m$
white	$aa\ A_r\ A_r\ BB\ A_p\ A_p\ A_m\ A_m$

In this A is the ground factor for colour, so that all aa forms are white; A_m a factor which in a recessive condition inhibits the formation of anthocyanin, so that $a_m\ a_m$ forms are pinkish white, while A_r, B and A_p are to be taken as intensifying and modifying factors. Besides these 5 factors there surely are still more, as in my material forms occur which cannot be explained with the aid of the factors mentioned. The analysis, however, has not yet been finished.

§ 5. *Historical survey of the appearance of flower colours*

In how far the colours will have originated in the same chronological order as the one I mentioned in § 1, cannot be stated with certainty. By using the oldest known data and the data known on the origin of new factors, however, it is possible to imagine the presumable previous history of the flower colours.

Of the two forms occurring first in literature: purple and white, purple is probably the older. This conception is not only supported by the fact that the wild pea bears purple flowers, but also that purple dominates over white, just as an ancestral form as a rule dominates over the derived form, as has been ascertained for by far the greater number of factor mutants. (BAUR, 7, 1930, p. 312). The period in which the factor mutation occurred of $A \rightarrow a$ I time at about 1000, as in the 16th century the white one already belonged to the most usual cultivated forms. In a corresponding manner rose can be considered a derived form of purple by factor mutation of $B \rightarrow b$. From the literary data (MILLER, 60, tome 6, p. 16) it may be concluded that the rose form appeared much later than the white one and I time the first appearance of rose flowers at about 1750.

The light purple forms differ in one factor from rose and violet. Very probably light purple will have originated through factor mutation $A_r \rightarrow a_r$ from rose. In favour of this is the fact that violet did not become known until later. I assume that about 1890 the light purple form appeared first.

From the cross with light purple through combination of factors violet first appeared in TEDIN's cultures about 1911.

Apple blossom differing from purple in one factor and behaving recessively with respect to purple, can doubtlessly be derived from it through factor mutation of $A_p \rightarrow a_p$. Apple blossom having been grown for a long time I time the first appearance at about 1850. From crossing apple blossom and rose there originated apple rose in 1929, while in 1930 by crossing apple blossom and violet apple violet arose.

The pinkish white form was first noticed by Dr. MANSHOLT in 1922. As discussed in § 3 this form also differs from purple in one factor, purple dominating. From this it was inferred that the pinkish white form will have originated through factor mutation of $A_m \rightarrow a_m$. As the appearance of a white form in a purple-flowering culture immediately strikes the eye and before 1922 nothing was known of the pinkish white form, I time its initial appearance at 1922.

Though the given representation can but give an approximate survey of the presumable history underlying the flower colours known

TABLE 10.   Historical survey

factor mutation		combination of factors	
	white ± 1000		
		red	1929
	rose ± 1750	dilute apple rose	1929
purple	apple blossom ± 1850	deep apple rose	1929
		apple violet	1930
	light purple ± 1890	violet	± 1911
	pinkish white 1922		

at the present time, I think it worth while composing a genealogical table, which may possibly be further verified.

§ 6. *Relation between flower colour and leaf axil colour*

It may be generally stated that flower and axil show corresponding colours. This is the case with purple, violet (pl. I, fig. 3, 5), rose (pl. II, fig. 4, 6) and light purple.

For the axil a ground factor for colour has been demonstrated. (TSCHERMAK, 87, 1912, p. 150).

From the cross of coloured flower with leaf axil colour × coloured flower without leaf axil colour TSCHERMAK deduced a monofactorial difference. This factor was called D. Moreover he assumed a factor C for leaf axil colour showing absolute coupling with the factor for flower colour A.

KAPPERT (49, 1923, p. 44) showed that with respect to the axil colour we may distinguish between a double ring in the leaf axil and a single ring, usually called double spot and single spot. This difference is based upon 1 factor. The TEDINS (85, 1926, p. 106) determined the existence of 3 allelomorphs of the factor D, of which D^w (double spot) dominates over D (single spot) and d (colourless), while D dominates over d.

Both D^w and D are to be taken as ground factors; if one of these multiple allelomorphs is present, the colour factors can manifest themselves. For simplicity's sake I shall only use D in the subjoined discussion, D^w having the same effect only with this difference that a double spot arises instead of a single spot.

As in DD forms of purple, violet, rose and light purple the flower colour always corresponds with the leaf axil colour a pleiotropic action of the factors A, A_r and B may be concluded.

KAPPERT (48, 1921, p. 204) brought forward the relation between flower colour and leaf axil colour of the pea and drew conclusions from this on the occurrence of absolute linkage in general. These conclusions, however, are founded in my opinion on an incorrect base. He pointed out that various cases of pleiotropic factors have been described, which afterwards appeared to be based on absolute or very strong linkage of factors. As an example he mentioned the connection between the flower colour and the leaf axil colour in the pea, in which case initially a factor was mentioned controlling both flower colour and leaf axil colour and later after there had been introduced into the investigation a form (*dd*), in which the properties occurred separately, the flower colour and leaf axil colour were taken as characters transmitted independently. This latter, however, does not bear upon the above question; we may only infer from it that a new factor for leaf axil colour had been determined. The initially determined connection, however, ought to be fully maintained. The question pleiotropic factor or absolute linkage of factors refers to the pleiotropic ground factor *A*, respectively to the absolute linkage of ground factors for flower colour and leaf axil colour, all other factors being not concerned with this.

The relation of flower colour and leaf axil colour can serve as a case in which 2 absolutely linked factors were assumed, it afterwards being ascertained that instead of this ·a pleiotropic factor exercised its influence both on the flower and on the leaf axil colour.

Though in this place I do not want to contest KAPPERT's opinion on the infrequent occurrence of pleiotropic factors, I wish to point out that the connection between flower colour and leaf axil colour cannot serve as an example of an initially incorrect conception of pleiotropism and subsequent correction by assuming 2 absolutely linked factors in its stead.

KAPPERT (51, 1925, p. 584) takes *D* as a pattern factor and is of opinion that the anthocyanin factors influence the action of the pattern factor. On the analogy of the action of the factor *A*, it is, however, desirable to take *D* as a ground factor, enabling the anthocyanin factors to manifest themselves in the leaf axil.

CORRENS (26, 1928, p. 164) brought forward in a more general connection the interaction of the discussed colour factors. He pointed out that both flower colour and leaf axil colour are due to the action of the same factors (A, A_r and B), but that besides this the place where they manifest themselves, should be determined. CORRENS assumed that they could be localized through 'Entwicklungsvorgänge' which have no concern with the nucleus.

From what has been said about the relation between flower colour and axil colour, it appears that the ground factor for leaf axil colour has been demonstrated, and that the locating action may be attributed to this factor *D*. From this it follows that the spot where the colour is manifested is factorially fixed.

STERN (75, 1930, p. 19) points out in discussing the three multiple allelomorphs D^w, D and d that in the homozygous forms of this series the total quantity of pigment 'derart verringert wird, dasz weniger Gebiete die Farbe ausbilden'. In my opinion this representation is not correct; the place is primarily determined by one of these factors and if the colour factors are present, pigment will be formed in this place.

Both the new flower colours apple blossom and pinkish white deviate from the rule that in the presence of the factor D flower and leaf axil show a corresponding colour. For the leaf axil colour of the apple blossom plants is purple. In the F_2 of crosses with purple flowered plants, therefore, there only occurs a segregation according to the flower colours, all plants possessing purple leaf axils. Hence it follows that the above discussed factor for apple blossom should be taken as a factor working only in the flower. In accordance with this is the fact that the apple rose plants have a leaf axil colour as the rose ones and the apple violet plants have a leaf axil colour as the violet ones. As both apple blossom and its derived forms apple rose and apple violet are hard to distinguish in a juvenile stage, the colour of the leaf axil forms a distinct criterion.

The leaf axil colour of the pinkish white plants is dull rose and easily distinguishable from those of the light purple or rose plants. Pinkish white colour of the flower and dull rose leaf axil colour always go together. The question may be put whether it is really one and the same factor that controls the pinkish white flower colour and the dull rose leaf axil colour. A factor having a different effect in the flower and in the leaf axils of the pea has not been previously described and though it can very well be imagined we have to take into account the possibility of absolute linkage between a factor for pinkish white flower colour and a factor for dull rose leaf axil colour. As long, however, as the type is maintained as such and the supposed linkage is not broken, I take A_m as a factor that has a different effect on the flower and on the leaf axil.

To give a survey of the flower colours and the appertaining leaf axil colours, I subjoin a scheme in which are given all phenotypes theoretically possible, possessing besides the ground factors A and D one or more colour factors. All aa forms have no colour in flower and in leaf axils, all $AA\ dd$ forms have a colourless leaf axil.

In the above we distinguished 6 factors, 3 of which (A, A_r, B)

TABLE 11.   Survey of the formulas

No.	genotype	flower	leaf axil
1	$AA\ BB\ A_r\ A_r\ A_m\ A_m\ A_p\ A_p\ DD$	purple	purple
2	$AA\ BB\ A_r\ A_r\ A_m\ A_m\ a_p\ a_p\ DD$	apple blossom	purple
3	$AA\ BB\ A_r\ A_r^{\,z}\ a_m\ a_m\ A_p\ A_p\ DD$	pinkish white	dull rose
4	$AA\ BB\ A_r\ A_r\ a_m\ a_m\ a_p\ a_p\ DD$	pinkish white	dull rose
5	$AA\ BB\ a_r\ a_r\ A_m\ A_m\ A_p\ A_p\ DD$	violet	violet
6	$AA\ BB\ a_r\ a_r\ A_m\ A_m\ a_p\ a_p\ DD$	apple violet	violet
7	$AA\ BB\ a_r\ a_r\ a_m\ a_m\ A_p\ A_p\ DD$		
8	$AA\ BB\ a_r\ a_r\ a_m\ a_m\ a_p\ a_p\ DD$		
9	$AA\ bb\ \ A_r\ A_r\ A_m\ A_m\ A_p\ A_p\ DD$	rose	rose
10	$AA\ bb\ \ A_r\ A_r\ A_m\ A_m\ a_p\ a_p\ DD$	apple rose	rose
11	$AA\ bb\ \ A_r\ A_r\ a_m\ a_m\ A_p\ A_p\ DD$		
12	$AA\ bb\ \ A_r\ A_r\ a_m\ a_m\ a_p\ a_p\ DD$		
13	$AA\ bb\ \ a_r\ a_r\ A_m\ A_m\ A_p\ A_p\ DD$	light purple	light purple
14	$AA\ bb\ \ a_r\ a_r\ A_m\ A_m\ a_p\ a_p\ DD$		
15	$AA\ bb\ \cdot\ a_r\ a_r\ a_m\ a_m\ A_p\ A_p\ DD$		
16	$AA\ bb\ \ a_r\ a_r\ a_m\ a_m\ a_p\ a_p\ DD$		

have a pleiotropic effect, while 1 (A_p) acts only in the flower, 1 (A_m) has a different effect in flower and leaf axil and 1 (D, respectively D^w) acts only in the leaf axil.

Of these 7 forms have not yet been observed, 1 of which is recessive for a_r, b and a_p, 6 being recessive for a_m. On the analogy of the data on the crosses of apple blossom with other forms it may be inferred that the former will be apple light purple, whereas the $a_m\ a_m$ forms just as the well-known form in which A_m inhibits anthocyanin-formation in a recessive condition, will be pinkish white. Presumably, however, the types will be distinguishable with the aid of the leaf axil colour.

TABLE 12.   Purple (colourless) × apple (single spot)

year	No.	purple		apple		total
		axilcolour	colourless	axilcolour	colourless	
1928	303	93	31	32	9	165

theor. 9 : 3 : 3 : 1

		92.8	:	30.9	:	30.9	:	10.3
	m	6.37		5.01		4.99		3.12
	D/m	0.031		0.020		0.220		0.416

As to the localization of D with respect to the colour factors it is known that D is transmitted independently with respect to A, A_r and B (H. and O. TEDIN, 86, 1928, p. 57). On the relation $D — A_m$ I possess no data, while independent transmission of D and A_p could be ascertained, as appears from the table 12.

§ 7. *Relation between flower colour and seedcoat colour*

MENDEL (59, 1866, p. 14) already pointed to a coincidence of coloured flower and coloured seedcoat: the purple flowered plants had a grey testa, wheras the white flowered plants had a colourless testa. This going together was afterwards attributed to the pleiotropic action of the ground factor for colour A. (the TEDINS and WELLEN-SIEK, 84, 1925, p. 533).

In the case of A_r and B it was demonstrated that they affect the seed colour in a similar way as these factors act in the flower. (H. and O. TEDIN, 86, 1928, p. 3). In my research it appeared that A_p and A_p have no effect on the colour of the seedcoat. Conversely it is not known of the factors Pl, M, F, Oh, Z, Mp for the colour of the seedcoat that they affect the flower colour (H. and O. TEDIN, 86, 1928, p. 4).

From this it follows that by the side of pleiotropic factors, there exist others that only affect the flower colour or only the seedcoat colour.

§ 8. *Relation between the flower colour factors and the leaf colour in spring*

Of various forms the anthocyanin formation in the leaf epidermis was traced during spring. The colour first appears in the cells round the stomata, while in cases, in which more connected cells show colour the cells round the stomata are coloured most.

The purple forms show a purple colour in the leaf epidermis. The various purple forms, however, vary in the degree of anthocyanin formation. The Fletumer rozijnerwt (raisin pea) shows little pigment, others on the other hand may possess a great deal of pigment.

The purple flowered Solo plants without axil colour (dd) only show a purple margin 2 to 3 cells wide around the leaf; this should probably be connected with the dd constitution.

I was struck by the fact that the coloured leaf epidermis cells of the rose plants show the same colour as those of the purple plants.

From this it may be inferred that the colour formation in the leaf epidermis occurs independently of the factor *b*. On the other hand a connection could be ascertained with the factor a_m; for all pinkish white forms show a dull rose colour in the leaf epidermis.

The degree of anthocyanin formation may be independent of the colour factors; that the colour that arises is connected with part of the flower colour factors occurring in the plant, however, is certain.

§ 9. *The flower colour of the closely related forms*

The above forms are all classed among the species *Pisum sativum L*, ampl. ASCH. & GR., to which both *Pisum arvense L.* and *Pisum sativum L.* belong.

I am much indebted to Professor VAVILOV for sending me some new forms. According to GOVOROV's description (36, 1928, p. 517) these forms deviate in flower colour from those examined. He supposed that the peculiar dirtish pink or creamy hue of the standard is determined by the presence, beside anthocyanin, of a special pigment belonging to the group of flavones. These forms were described as subsp. *asiaticum* of *Pisum sativum L.*, ampl. ASCH. & GR.

Of the other species little is known concerning the genetic constitution of the flower colour.

Pisum elatius has purple flowers of a darker hue, however, than the purple colour of *Pisum sativum*. In crossings with white forms the difference appeared to be monofactorial.

Pisum fulvum has creamy coloured flowers (Index Kewensis, 1895, p. 545).

SUTTON (78, 1911, p. 365) used for crossing a self coloured magenta form which he found growing wild in Palestine and gathered there. This form resembles much *Pisum quadratum*. From the research it appeared that the F_1 of magenta × white (*P. sativum*) is purple. Because the F_1 is sterile to a strong degree, the genetic constitution of the magenta flower colour could not be further examined. To SUTTON's paper coloured plates are appended; from these, however, it cannot be concluded whether the magenta type shows correspondence with the above described apple blossom type, which is likewise self coloured. Messrs. SUTTON were kind enough to give me the magenta coloured Palestine pea for further investigation. I wish to express my sincere thanks for their kindness. Owing to the sterility of the F_1's which renders it impossible to carry out a factor analysis, I intend to cross the magenta form with various flower colours and to conclude from the F_1's something about the hereditary difference between magenta and other flower colours.

When this publication was in the press there appeared a preliminary communication by FEDOTOV (33, 1930, p. 523) on the genetic constitution of some new flower colours, which are called bright red (crimson), light crimson, creamy purple, creamy purple pale, creamy rose, creamy white, purple with dark wings and marked veins on the standard and bluish white. To explain the genetic behaviour 4 new factors are added to the known A, A_r and B. Among these C_r, C_v and B_1 have an intensifying effect, while C_m determines the presence of cream colour in the flowers. C_m is independent of the fundamental factor for flower colour, A. From this FEDOTOV concluded that the determination of A should be rectified and it should be regarded as fundamental factor for anthocyan colour of the flower, not for flower colour in general, as it was regarded up to now.

CHAPTER II

PATCHED AND DOTTED FLOWERS

Section 1. Review of the literature on the heredity of the mosaic flowered plants in general

The heredity in mosaic flowered plants is among the most difficult problems as appears from the many explanations given about it.

A great number of mosaic flowered plants are mentioned in literature. The communications dated before 1900 are mostly descriptions, yet in various publications views are given on the origin of flower mosaic.

In BRAUN's (8, 1851, p. 336) opinion the same flower colour varieties as exist in *Mirabilis*, can also occur in one and the same mosaic plant, so that hereditarily different parts arise in a plant.

A similar explanation NAUDIN (62, 1863, p. 192) gave.

DE VILMORIN (89, 1852, p. 9) assumed that striped flowers occur only in species which are themselves coloured, but which possess a white or yellow variety and that in the white (or yellow) cultures plants may occur with narrow stripes, which become the starting-point of a mosaic flowered population.

DARWIN (27, 1868, 2, p. 37) inferred from DE VILMORIN's data, that striped plants revert to the original type, to characters lost by variation. He pointed out (27, 1868, 2, p. 70) that the striped plants have a latent tendency to become unicoloured.

From the above it appears that the deviating behaviour of the mosaic flowered plants was already observed before 1900.

In his Mutationstheorie (90, 1901—1903) DE VRIES communicated a great number of observations on the mosaic flowered plants. He made a special study of *Delphinium*, *Antirrhinum*, *Hesperis* and *Clarkia*. DE VRIES (90, 1903, p. 514) considered the flower mosaic a pattern, in which, therefore, the patches are due to differentiation, all cells having the same genetic constitution. The factor for striping is due to mutation of a flower colour factor. The effect of this factor is strongly influenced by external circumstances, resulting in a variation in the degree of flower mosaic. The factor for striping appeared to be recessive with respect to the factor for unicoloured flowers. DE VRIES pointed out that if the same holds good for this factor as for others the homozygous recessive plants ought to breed true for striping. This, however, not being the case, DE VRIES inferred that the hereditary behaviour cannot be explained by a simple mendelian factor.

CORRENS (18, 1902, p. 594) initially tried to explain the genetic behaviour of mosaic flowered *Mirabilis* forms by assuming that the mosaic plants were heterozygous, having originated from the cross of 2 unicoloured types and that the mosaic character was due to shifting dominance. In a subsequent communication he regarded the flower mosaic as a pattern, formed by a complicated action of 2 factors (19, 1905, p. 70). In his later communications (20, 1909, p. 291 and 21, 1910, p. 418) CORRENS also considered the various parts of the mosaic flowers as hereditarily identical. The occurrence of unicoloured plants in the progeny of mosaic he explained by assuming that the self colour gene became separated from the mosaic gene.

WHELDALE (94, 1909, p. 1) ascertained that the factor for the mosaic flower of *Antirrhinum* is allelomorphic with the colour factors $D - d$. She regards the flower mosaic as a pattern caused by the said allelomorph.

GREGORY (37, 1910, p. 123) examined mosaic flowered forms of *Primula* sinensis and determined a factor for striping which is multiple allelomorphic with the colour factors $Y - y$.

BAUR (4, 1911, p. 205) like WHELDALE investigated mosaic flowered *Antirrhinums*. He regarded the factor for striping as multiple allelomorphic with 2 of the well-known allelomorphic colour factors and ascertained that this factor for striping is ever-mutating. In his subsequent publication (5, 1914, p. 304) BAUR emphasized the fact that the flower mosaic is not due to differentiation of a pattern, but that the patched flower is a mosaic of genetically different parts. In his subsequent communications BAUR also maintained the explanation of an ever-mutating factor which can mutate in the somatic cells throughout the development.

PUNNETT (66, 1922, p. 255) investigated mosaic flowered *Lathyrus* forms. He left it undecided whether a factor for striping exists or the mosaic is due to a special action of the colour factors in a heterozygous condition.

BATESON regarded flower mosaic as a pattern. In a subsequent publication (3, 1926, p. 230), however, he deemed it possible that through segregation of the gene quantities in the somatic cells genetical differences arise.

IMAI (43, 1925, p. 46; 44, 1927, p. 255) mentioned in the case of *Pharbitis* the occurrence of a cream flower with a few magenta coloured fine splashes. This behaves as a simple recessive to the self coloured magenta and it produces some self coloured mutants in its progeny.

DEMEREC (30, 1926, p. 35) determined for the mosaic flowered *Delphinium Ajacis* 3 multimutating genes. He regarded the spotted flower as a mosaic of genetically different parts.

CLAUSEN (14, 1926, p. 25) mentioned mosaic flowered *Viola* plants; he explained these by assuming somatic mutation of the colour factor $L \rightarrow l$. So he regarded the spots as genetically different from the other parts.

CHITTENDEN (13, 1927, p. 416) gave an absolutely different explanation and pointed out that many of the published cases may be more easily interpreted on the basis of a plastid inheritance.

IKENO (42, 1928, p. 189) explained the heredity of the mosaic flowered *Portulaca grandiflora*, the mechanism of which is very difficult to understand. IKENO's views come to this: the self coloured form is pure for 'Farbgenen' and the white form is pure for 'Weissgenen', while the gametes, which produce mosaic, contain both 'Farb'- and 'Weissgenen'. According as more 'Farbgenen' are present, the more pronounced is the mosaic character of the flowers.

CAYLEY (10, 1928, p. 529) determined that the flames of 'broken' tulips can be transmitted by inoculation and may therefore be regarded as contagious.

EYSTER (32, 1928, p. 670) studied the flower mosaic of *Verbena*. He explained the flower mosaic by assuming quantitative differences of a colour factor based upon the hypothesis that the ratio of the 2 different species of genomeres in one gene may be different.

CLAUSEN (16, 1930, p. 360) mentions in *Nicotiana* carmine coral striped flowers. He assumes that the flower mosaic is based upon sporadic loss of a chromosome fragment during development and points out that it is not impossible that variegation phenomena are generally due to some form of chromosomal instability.

KOJIMA (55, 1930, p. 328) mentions scarlet and yellow mosaic coloured flowers of *Celosia cristata*. From his experiments he concluded that the factor *a* for yellow is unstable and the mutation *a* → *A* frequently takes place. This new factor *A* for scarlet is quite stable.

From the above enumeration it appears that a great number of cases of mosaic flowers have been genetically examined, but at the same time that a great many explanations of the heredity have been given. Though it is quite possible that different cases of flower mosaic are genetically different, it is very doubtful whether there really exist so many differences as may be concluded from this enumeration.

Section 2.   Self-pollinations

§ 1. *Introduction*

The material for the research on the heredity of the purple patched (mosaic) [1]) flower colour was received from Dr. R. J. MANSHOLT of Westpolder (Groningen). In the summer of 1923 Dr. R. J. MANSHOLT observed 1 plant with purple patched flowers in a culture of purple flowering Fletumer rozijnerwten. Since then a similar departure in the purple cultures has never occurred again. The purple patched plants [2]) cultivated by me all belong to the progeny of this one plant. In the summer of 1926 I received two flowers of a purple patched plant for making crossings, while in the autumn of that year Dr. MANSHOLT was kind enough to send me some plants of the purple patched culture together with some seeds of the Fletumer rozijnerwt, the form from which the purple patched forms originated.

Dr. MANSHOLT informed me that the purple patched form was not to be got constant and that in the progeny by the side of purple patched plants there also occurred purple, purple dotted and nearly white

---

1) Because patched flowers as to their pattern show a great resemblance to variegated leaves the patched flowers are also called variegated. It is, however, undesirable to use the same name for fundamentally different characters, so that it is essential to adopt CHITTENDEN's terminology (13, 1927, p. 408) and call irregular leaf colour affecting chorophyll variegated, patched flowers mosaic.

2) For the sake of brevity we shall speak in the following pages of patched plants or patched branches, when patched flowered plants, respectively, branches are meant.

forms, that is flowers with a slight purple patching or purple dotting.

He thought it most likely that the new form had arisen from the purple flowered culture. This supposition is probably true, since for the rest the phenotype of the two forms is perfectly conform. Both types flower very early, and the genotypical constitution for the length factors also appeared to be the same (*le le La La lb lb*).

§ 2. *Description of the mosaic flower*

The mosaic flowers are purple patched on a white background (plate II, fig. 9). In one and the same plant they differ more or less in the degree of patching, the patches also varying in size. In every case there is a clearly marked separation of the purple patches from the surrounding colourless cells. The patches on the lighter coloured standard are often difficult to distinguish, the more striking, however, are the purple patches on the in- and outside of the wings. On the purple patched plants normal purple flowers are manifold, sometimes a purple patched branch passes into purple at the top, completely purple branches being very frequent. The purple patched plants vary in the degree of patching. Some plants were initially recorded as 'white', while later a sporadic purple patch was observed on one of the flowers. Other plants on the other hand showed a strong purple patching, so that a number of entirely purple flowers occurred on them.

The flower colour of the pea occurs in the epidermal cells (fig. 1, top to the left). From this it follows that when the genetic constitution of the epidermis does not always correspond to that of the subepidermis, from which the gametes are formed, the progeny cannot be foretold from the flower colour. This is actually the case here. It is, therefore, a happy coincidence that besides epidermically in the flower the patching is also exhibited subepidermically in the leaf axil (fig. 1, top to the right).

Besides purple branches, purple flowers and purple flower patches, the purple patched plants also show purple dotted parts. Sometimes a whole branch is purple dotted, sometimes the purple patched branch passes into purple dotted, sometimes the flowers are both purple dotted and purple patched. The dots (fig. 1, bottom to the right) which, like the spots, are epidermic, differ in intensity from the purple patches; they are of a lighter purple. Neither does the marked transition from colourless to coloured, characteristic of purple patched occur. The dot often consists of some cells, of which the central one

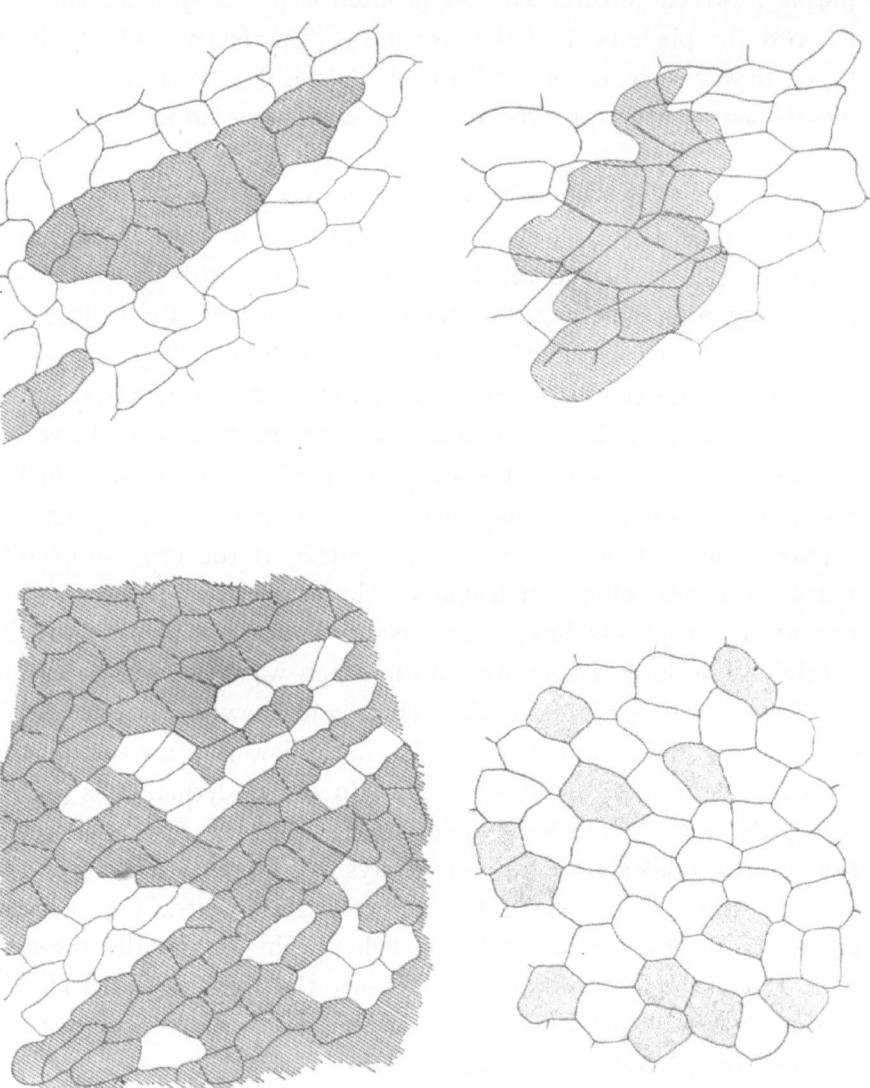

Fig. 1.   Top to the left: epidermis of a patched flower.
Top to the right: epidermis and subepidermis of a patched plant.
Bottom to the left: epidermis of a crypto purple flower.
Bottom to the right: epidermis of a purple dotted flower.
E. = 200 ×.

is of a darker shade than the surrounding ones. In flowers which
show both patterns, colourless dots appear in the purple patches

(fig. 2), which but for their being colourless, remind us strongly of the light purple dots of the dotted pattern.

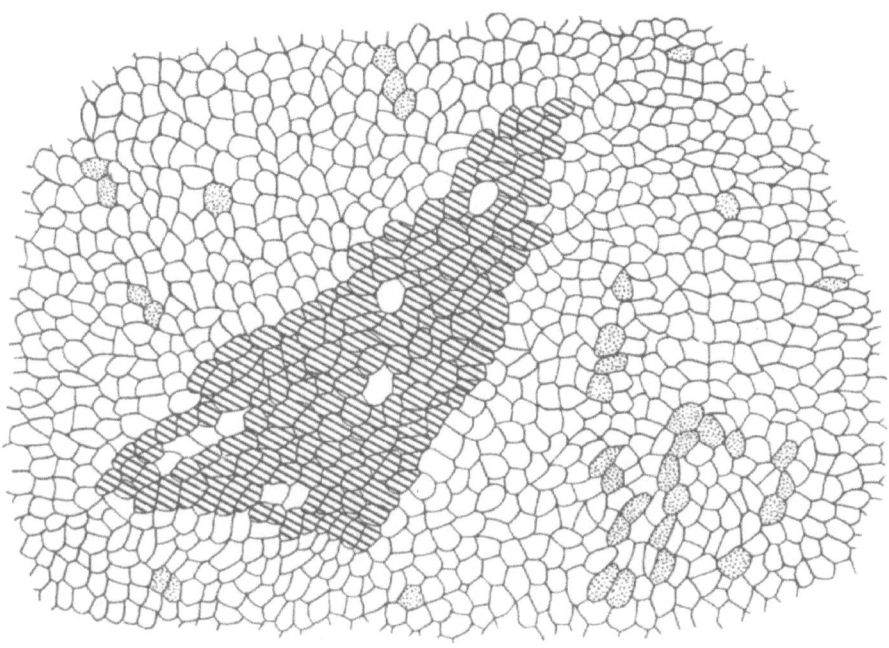

Fig. 2. Epidermis of a patched + dotted flower.
E. = 100 ×.

§ 3. *Progeny of the purple patched plants*

For 4 years purple patched plants have been propagated. The data appear in the subjoined table 13.

From the table it appears that in all these cases purple patched yields a mixed progeny and that in many cases purple plants occur in it, the percentage of which is strongly variable. On the whole heavily patched plants attain the highest percentage in the progeny. Nearly always purple dotted plants appear in the progeny of purple patched forms. The percentage also varies greatly, on the whole it is higher according as the flower patching is slighter.

I have not succeeded in finding patched plants constant for purple patching; when the plants with stronger patching were propagated a purple progeny appeared by the side of a purple patched one, when the plants of a slighter purple patching were propagated, purple dotted forms occurred by the side of purple patched forms. So I can corroborate Dr. MANSHOLT's report.

TABLE 13. Self-pollination of purple patched plants

year	No.		purple	purple dotted	purple patched	purp. dotted + purple patched	total
1927	27	not branched		9	13	22	22
,,	71	,,	3	6	18	24	27
,,	72	,,	6	7	14	21	27
1928	27.4	,,	7	30	13	43	50
,,	27.8	,,		19	14	33	33
,,	27.9	,,	4	3	17	20	24
,,	27.13	,,		18	2	20	20
,,	27.14	,,	6	15	25	40	46
,,	27.19	,,		25	12	37	37
,,	27.21	,,	4	21	26	47	51
,,	71.3	,,	11	7	8	15	26
,,	71.4	,, .	15	3	25	28	43
,,	71.18	,,		5	7	12	12
1929	150	branch 1		5	8	13	13
,,	,,	,, 2		12	23	35	35
,,	151	,, 1	1	1	12	13	14
,,	,,	,, 2	5	2	16	18	23
,,	152	not branched	1	22	5	27	28
,,	153	,,	2	12	8	20	22
,,	154	,,		26	8	34	34
,,	155	branch 1	1	1	13	14	15
,,	,,	,, 2	1	2	5	7	8
,,	,,	,, 3	6	1	13	14	20
,,	,,	,, 4	2	8	4	12	14
,,	,,	,, 5	1		6	6	7
,,	,,	,, 6	3	1	4	5	8
,,	156	,, 1	3	4	12	16	19
,,	,,	,, 2	1	19	4	23	24
,,	,,	,, 3	5	9	14	23	28
,,	,,	,, 4	4	7	21	28	32
,,	,,	,, 5	4	1	13	14	18
,,	,,	,, 6	1	6	7	13	14
,,	,,	,, 7	1	10	8	18	19
,,	157	,, 1	14		1	1	15
,,	,,	,, 2	10	2	4	6	16
,,	,,	,, 3	6	7	20	27	33
,,	,,	,, 4	15		4	4	19
,,	,,	,, 5	2	8	8	16	18
1930	131	not branched	5	11	39	50	55
,,	132	,,		3	18	21	21
,,	134	,,	1	8	13	21	22
,,	135	,,	7	17	31	48	55
,,	138	branch 1		3	7	10	10
,,	,,	,, 2	2	7	24	31	33

Below the observations are recorded made on the progeny of the different branches of the purple patched plant and on the progeny of the different forms originated from purple patched.

§ 4. *Variously coloured branches on purple patched*

The progeny of the purple branches consisted of purple, purple dotted and purple patched plants. If purple dotted + purple patched are considered 1 group (see § 15) a segregation can be ascertained in the progeny of the purple branches in 3 purple : 1 (purple dotted + purple patched), as appears from the following table.

TABLE 14. Self-pollinations of purple patched

1929	No.	branch	purple	purple dotted	purple pat-ched	purple dotted + purple patched	total	theor. 3 : 1		m	D/m
purple	129	1	54	3	21	24	78	58.5	19.5	3.67	1.226
,,	,,	2	39		11	11	50	37.5	12.5	3.12	0.481
purple patched	,,	3	2	4	24	28	30				
,,	,,	4		8	16	24	24				
,,	,,	5	6	2	43	45	51				
purple	130	1	37	4	11	15	52	39	13	3.04	0.658
purple patched	,,	2	3	3	26	29	32				
,,	,,	3		3	21	24	24				
purple	129a	1	45		11	11	56	42	14	3.35	0.895
,,	,,	2	37	6	9	15	52	39	13	3.04	0.658
purple patched	,,	3	3	3	23	26	29				
,,	,,	4		6	35	41	41				

From these observations it follows that the purple branches were heterozygous for the factor for purple, while the purple patched branches gave a similar progeny as the purple patched forms in table 13. The heterozygous purple branches were interpreted by the assumption of early mutation of a gene for (purple dotted + purple patched) to the factor for purple.

With respect to the purple flower patches the same may be assumed. According as the mutation occurs in a younger stage — the cell, therefore, in which the mutation takes place, will have to undergo many divisions — the purple patch will be larger; if the mutation occurs very late, but a small purple patch will arise.

In all cases in which purple branches were examined they appeared to be heterozygous. Cases in which both allelomorphs for (purple dotted + purple patched) mutated to the factor for purple, therefore, are not known.

Repeatedly a plant was first described as purple, while it appeared later, that purple patched lateral branches occurred on it. On further examination it appeared that the lower internodes had a patched leaf axil. In such cases mutation had occurred in a young stage of the main stem.

§ 5. *The purple plants from the progeny of purple patched*

From Dr. MANSHOLT I also received the seed of a purple-flowering plant, observed in the progeny of purple patched. From this the following culture was made.

TABLE 15.   Self-pollination of a purple plant from the progeny of purple patched

year	No.	purple	purple patched	total
1927	70	24	7	31
theor. 3 : 1		23.25	7.75	
		m 2.45	D/m 0.306	

Though the number is slight a 3 : 1 segregation may be assumed. In order to trace this the culture was propagated. The observations occur in the following table 16.

Of the 18 plants 4 purple plants appeared to be constant, 9 purple plants segregated into purple, purple dotted and purple patched, 5 purple patched plants giving a corresponding progeny to that of the plants mentioned in table 13. In the progeny of the 9 segregating purple plants a 3 : 1 ratio may be established, if purple dotted and purple patched are considered one group. The 9 segregating purple plants, therefore, were heterozygous for the factor for purple and the one for (purple dotted + purple patched).

Besides it appears from table 16 that from purple patched constant purple may arise. These purple plants were used in crossings and it

TABLE 16. Self-pollinations of the plants occurring in table 15

1927	1928								
flower colour	No.	purple	purple dotted	purple patched	purple dotted + purp. patched	total	theor. 3 : 1	m	D/m
purple	70.16	31				31			
,,	70.24	26				26			
,,	70.3	53				53			
,,	70.4	34				34			
purple	24.2	27	3	5	8	35	26.25 8.75	2.59	0.289
,,	70.13	20	1	7	8	28	21 7	2.23	0.448
,,	70.15	39	2	10	12	51	38.25 12.75	3.12	0.240
,,	70.5	42	10	3	13	55	41.25 13.75	3.24	0.231
,,	70.17	42	9	6	15	57	42.75 14.25	3.24	0.231
,,	70.2	45	12	2	14	59	44.25 14.75	3.35	0.224
,,	70.8	14	9	1	10	24	18 6	1.87	2.139
,,	70.9	31	4	8	12	43	32.25 10.75	2.78	0.450
,,	70.10	18	3	3	6	24	18 6	2.12	0
purple patched	70.18	2	12	7	19	21			
,,	70.22		15	14	29	29			
,,	70.30	1	3	14	17	18			
,,	70.7		15	1	16	16			
,,	70.11	1	11	4	15	16			

was ascertained from them that these purple plants are identical with the purple plants that do not descend from purple patched.

§ 6. *Purple dotted in the progeny of purple patched*

From table 13 it appears that the percentage of purple dotted varies greatly, that, however, purple dotted plants are very frequent.

A number of purple dotted plants due to self-pollination of purple patched appeared to be constant. This may appear from the subjoined summary.

In these cultures purple patched has never again been observed, while in crosses purple dotted also behaves as a constant form. Nor do somatic differences occur in one and the same plant. Purple dotted may, therefore, be considered a dotted colour pattern, all constituent parts of which are genetically identical.

TABLE 17.   Self-pollination purple dotted

year	No.	purple dotted
1927	71a	25
,,	311	49
,,	319	29
1929	136	34
,,	137	30
,,	138	35
,,	139	33
,,	140	34
,,	141	35
,,	142	35
,,	142a	35
1930	138a	86
,,	137	43
total		503

§ 7. *The 'white' plants*

In the description of the purple patched forms it has been pointed out that the pattern may vary greatly, that is from almost white ('white') to entirely purple.

On the whole it may be said that the different branches of one plant present about the same type of patching. In the progeny of 'white' plants many 'white' forms occur, while in the progeny of strong patching many plants with strong patching occur.

Here the difficulty presents itself that from the leaf axil colour it cannot be concluded whether the constitution of the subepidermal layer shows a strong or less strong patching or whether it is dotted, while the constitution of the epidermis, as has been said at the beginning of this chapter, sometimes departs from that of the subepidermal layer, from which the gametes are formed. This appears from the following tables, in which the progeny has been recorded of 'white' descending from purple patched cultures.

From this it appears that the progeny of the 'white' culture of plant 143 mainly consists of 'white' plants. On the other hand the progeny of the 'white' 144 contains a considerably larger number of

TABLE 18.   Self-pollination 'white' plants

year	No.	branch	purple	purple dotted	purple patched	'white'	total
1929	144	1			2	13	15
,,	,,	2		2	17	11	30
,,	,,	3	1	1	16	4	22
,,	143	1		1		19	20
,,	,,	2		4	2	29	35
,,	,,	3		1		13	14

purple patched ones. The first I think I may regard as normal, while in the 2nd case the 'white' epidermis probably differed genotypically from the subepidermis which presumably showed a stronger degree of patching.

The relation between purple patched, purple dotted and purple will be further discussed after the crosses, because the hypothesis I set up, is based on the difference in behaviour of purple patched in reciprocal crosses.

Section 3.   The crosses
Between the purple patched and the various forms discussed in chapter I, several crossings have been made.

§ 8.   *Purple patched × white*
The F_1 of purple patched × white was not uniform and consisted like the self-pollinations of purple patched of purple, purple dotted and purple patched, as appears from the subjoined table.

TABLE 19. F_1 purple patched × white

year	No.	purple	purple dotted	purple patched	total
1928	143		2	3	5
,,	152-153	1	6	1	8
,,	211	2	1	2	5
total		3	9	6	18

From the above purple dotted F_1 plants 7 F_2's were cultivated. The observations occur in the subjoined table.

TABLE 20. F_2 purple patched × white

year	No.	purple dotted	purple patched	purple dotted + purp. patched	white	total	theor. 3 : 1		m	D/m
1929	104	34		34	11	45	33.75	11.25	2.91	0.086
„	105	42	1	43	16	59	44.25	14.75	3.27	0.382
„	106	59	1	60	18	78	58.50	19.50	3.87	0.388
1930	72	30	2	32	9	41	30.75	10.25	2.82	0.443
„	73	21		21	8	29	21.75	7.25	2.29	0.328
„	74	40		40	15	55	41.25	13.75	3.16	0.395
„	75	14		14	3	17	12.75	4.25	1.87	0.668
„	76	53		53	19	72	54	18	3.64	0.274

From this follows a segregation into 3 purple dotted : 1 white, some purple patched plants occurring.

From 1 of the purple patched F_2 plants from the cross purple patched × white an F_3 was cultivated.

TABLE 21. F_3 from one purple patched F_2 plant from purple patched × white

year	No.	branch	purple	purple dotted	purple patched	purple dotted + purple patched	total
1930	122	1	6	2	29	31	37
„	„	2	2	8	22	30	32
„	„	3		3	15	18	18

§ 9.   *Purple patched × coloured*

The F_1 plants of the crosses purple patched × coloured were always purple. This was established for purple ($AA\ A_r\ A_r\ BB\ A_m\ A_m\ A_p\ A_p$), light purple ($AA\ a_r\ a_r\ bb\ A_m\ A_m\ A_p\ A_p$), pinkish white ($AA\ A_r\ A_r\ BB\ a_m\ a_m\ A_p\ A_p$), apple ($AA\ A_r\ A_r\ BB\ A_m\ A_m\ a_p\ a_p$) and rose ($AA\ A_r\ A_r\ bb\ A_m\ A_m\ A_p\ A_p$). From the occurrence of purple F_1

TABLE 22. F_2 purple patched × apple

year	No.	purple	purple dotted	apple	apple dotted	total	theor. 3 : 1 resp. 9 : 3 : 3 : 1	m	D/m
1929	117	28		8		36	27 9	2.64	0.379
,,	118	54		18		72	54 18	3.67	0
,,	120	31		9		40	30 10	2.78	0.360
,,	116	32	10	12	3	57	32.04 10.68 10.68 3.56	3.74 2.97 2.90 1.83	0.011 0.229 0.455 0.306
,,	119	43	15	15	4	77	43.29 14.43 14.43 4.81	4.33 3.41 3.41 2.13	0.067 0.167 0.167 0.380
,,	121	24	8	6	3	41	23.04 7.68 7.68 2.56	3.24 2.48 2.56 1.54	0.296 0.129 0.656 0.286

TABLE 23. F_2 purple patched × rose

year	No.	purple	purple dotted	rose	rose dotted	purple patched	total	theor. 3 : 1 resp. 9 : 3 : 3 : 1	m	D/m
1929	107	98		34			132	99 33	4.94	0.202
,,	111	56		20			76	57 19	3.74	0.267
,,	108	137	47	45	14		243	136.71 45.57 45.57 15.19	7.74 6.06 6.09 3.78	0.037 0.236 0.093 0.315
,,	109	36	11	13	3		63	35.43 11.81 11.81 3.94	3.97 3.12 3.06 1.93	0.144 0.259 0.389 0.487
,,	110	87	31	30	10		158	88.87 29.62 29.62 9.87	6.17 4.88 4.89 3.04	0.303 0.283 0.077 0.043
1930	62	117	41	38	14	2	212	119.25 39.75 39.75 13.25	7.15 5.63 5.71 3.52	0.315 0.222 0.307 0.213

TABLE 24. F$_3$ from a purple patched plant from the cross purple patched × rose

No.	branch	purple	rose	purple dotted	rose dotted	purple patched	rose patched	purple dotted + purple patched	rose dotted + rose patched	total	theor. 3 : 1 resp. 9 : 3 : 3 : 1	m	D/m
121	1			14	4			14	4	18	13.5 4.5	1.87	0.267
,,	2			17	6			17	6	23	17.25 5.75	2.06	0.121
,,	3			19	5	3		22	5	27	20.25 6.75	2.34	0.748
,,	4		2	14	4			14	4	20	15 5	1.87	0.534
,,	5			27	10			27	10	37	27.75 9.25	2.59	0.289
,,	6			48	14			48	14	62	46.5 15.5	3.46	0.433
,,	7			10	2	2	3	12	5	17	12.75 4.25	1.73	0.433
,,	8			3	1			3	1	4	3 1	0.86	0
,,	9	2		27	8		2	27	10	39	29.25 9.75	2.69	0.093
,,	10	13	3	3	2			3	2	21	11.79 3.93 3.93 1.31	2.38 1.83 1.83 1.09	0.509 0.508 0.508 0.633
,,	11	29	10	7	3			7	3	49	27.56 9.18 9.18 3.06	3.56 2.70 2.81 1.69	0.404 0.304 0.776 0.036
,,	12	24	9	7	3			7	3	43	24.18 8.06 8.06 2.69	3.24 2.52 2.59 1.58	0.373 0.055 0.409 0.196
total										360			

plants in all these cases it might be concluded on the analogy of other crosses, that purple patched possesses the flower colour factors A_r, B, A_m and A_p.

Of the above crosses some F_2's were cultivated.

From table 22 it follows that in 3 cases a dihybrid segregation occurred and in 3 cases a monohybrid segregation. In the latter cases, therefore, the gamete from the purple patched form had the constitution A A_p. In the F_2 a new form appeared, which was called apple dotted (plate II, fig. 7), and which was recessive for the factor for dotted and the factor A_p.

In none of these F_2's a patched plant was observed, it did, however, occur in the subjoined F_2 of the cross purple patched × rose.

In the above F_2 (table 23) rose dotted (plate II, fig. 8) occurs as a new form. The F_2 corresponds to a 9 : 3 : 3 : 1 segregation, 2 purple patched plants occurring.

From 2 purple patched F_2 plants from the cross purple patched × rose an F_3 was cultivated (tables 24 and 25). The progeny of one of these plants occurs in the table 24.

The branches 10, 11 and 12 were purple; their progeny segregated into purple, rose, purple dotted and rose dotted, while that of the remaining branches segregated into 3 (purple dotted + purple patched) : 1 (rose dotted + rose patched). From this it follows that the F_2 plant had the constitution B b. Branch 4 produced beside purple dotted and rose dotted some rose plants, branch 9 some purple plants, while for the rest some purple patched and some rose patched plants occurred. These latter presumably had the constitution bb and moreover the constitution for patching.

The other purple patched F_2 plant from the cross purple patched × rose had 2 branches with a purple leaf axil colour. Conform to this the progenies of branches 1 and 2 segregated into 3 purple : 1 (purple dotted + purple patched), while branches 3 and 4 did not give a definite ratio. The observations are given in the subjoined table 25.

For the rest F_3's were cultivated from rose dotted, of which it could be ascertained that they were constant. Rose branches or rose flower patches were not observed.

§ 10. *White × purple patched*

Besides crosses were made with purple patched as father plant.

TABLE 25. F₃ from a purple patched F₂ plant from the cross purple patched × rose

branch 1929		1930, No. 114							
	purple	purple dotted	purple patched	purple dotted + purple patched	total	theor. 3 : 1		m	D/m
1 purple	46	1	16	17	63	47.25	15.75	3.39	0.369
2 purple	31	1	10	11	42	31.50	10.5	2.78	0.179
3 purple patched	12	1	14	15	27				
4 purple patched	39	6	33	39	78				
total					210				

This reciprocal cross, however, was different, because purple patched F_1 plants never occurred. As this observation was very important with a view to the explanation of the heredity of the flower mosaic a great many crosses have been made. These occur in the subjoined table.

TABLE 26. F_1 white × purple patched

year	No.		purple	crypto purple	purple dotted	purple patched	total
1928	101—109	white × white flower on purple pat- ched plant			48		48
,,	110—128	,, ,, purple patched	4	1	79		84
,,	129	,, ,, ,,			4		4
,,	140—142	,, ,, ,,			7		7
,,	144—151	,, ,, ,,	17		28		45
,,	154—157	,, ,, ,,			10		10
,,	158—160	,, ,, ,,	2		9		11
,,	161—164	,, ,, ,,	2		18		20
,,	165—168	,, ,, ,,			16		16
,,	169	,, ,, ,,			3		3
,,	170—175	,, ,, ,,	1		36		37
,,	176—182	,, ,, ,,	16		27		43
,,	183—190	,, ,, ,,	3		47		50
,,	191—196	,, ,, ,,	5		14		19
,,	207—210	,, ,, ,,	3		10		13
,,	212—215	,, ,, purp. flower on purple patched	6		8		14
,,	216—222	,, ,, ,,	8		23		31
,,	223—226	,, ,, ,,	7		16		23
,,	227—228	,, ,, ,,	1		8		9
total			75	1	411	—	487

Of the F_1 plants mentioned in table 26, therefore, 75 were purple, 411 purple dotted, while 1 plant occurred which was not pure purple, but purple with white dots. This form was not observed before and designated crypto purple (plate II, fig. 11).

From the above crosses just as from the self-pollination of purple patched it may be concluded that purple patched has different gametes. On the analogy of other crosses it may be assumed that in the purple dotted F_1 plants the ground factor for colour A does not occur, the colour of all A forms being purple. In this connection most crosses were made with white as mother plant. White was chosen, because the dominating factor A does not occur in it, and moreover, because the white plant used will give many seeds, the number of seeds on the patched plant being small. White has green cotyledons, so that after crossing with patched, of which the cotyledons are yellow, it could be ascertained from the cross-seeds on account of their yellow cotyledons, that the crossing had succeeded.

The purple dotted form of the F_1 plants (plate II, fig. 10) differed from that due to self-pollination of the patched form (plate II, fig.12). The dotting on the F_1 plants was sometimes hard to distinguish and was limited to the margins of the wings. The homozygous purple dotted plants had much clearer markings and their entire wings were dotted. As in the F_2's the two types could not always be distinguished from each other with certainty, I have combined them in the following pages under the name purple dotted.

Though no purple patched F_1 plants occurred, yet several purple dotted plants were somatically not constant. Sometimes purple-flowering branches with purple dotted leaf axil colour appeared, while there also occurred branches with a purple leaf axil colour and purple dotted flower colour, sometimes both flowers and leaf axils being purple. Further branches were observed which were purple over the entire length for one half, for the other half they were purple dotted. In a few cases such a branch passed into purple at the top and sometimes into purple dotted.

§ 11. *The difference of the reciprocal F_1's*

From the crosses purple patched × white various conclusions may be drawn. Important to the research on the heredity of purple patched is the observation that the reciprocal crosses differ. From this it follows that something is involved that is not due to a gene localized in the nucleus, but that may most probably be attributed to the plasm. To bring about the flower mosaic the maternal plasm is evidently essential. In discussing the hypothesis of the heredity of the flower mosaic I shall revert to this (§ 15).

Further our attention should be directed to the fact that in the cross white × purple patched not a single purple patched plant occurred. From this it may be inferred that the factor which causes patched in the plasm of the purple patched plant, is manifested in the plasm of the white mother plant as purple dotted, for if the purple patched father plant were self-pollinated, purple patched plants would have occurred in the progeny.

§ 12. F_2 *white* × *purple patched*

As was communicated in § 10, the F_1 plants of white × purple patched were partly purple, partly purple dotted.

On being sown out the purple F_1 plants appeared to be heterozygous for the A factor, as appears from the subjoined table. In connection with what follows the colour of the leaf axil has also been given here.

TABLE 27. F_2 white × purple patched

1927, Fl 38a. 1		1928, F2 325		
leaf axilcolour	flower colour	purple	white	total
purple	purple	99	31	130

theor. 3 : 1   97.5 32.5

m 4.97   D/m 0.302

As white does not possess the A factor, it may be inferred that an A gamete must be due to purple patched. The white plants in the F_2 were pure white, not a single one was purple patched or purple dotted. The A gamete from purple patched may be regarded identical with that of purple plants that have no purple patched among their ancestors.

The purple dotted F_1 plants of the cross white × purple patched, which did not bear deviating branches, gave a segregation in the F_2 into 3 purple dotted : 1 white, as follows from the subjoined table.

TABLE 28. F_2 white × purple patched

1927, Fl 45.1		1928, F2 324		
leaf axil colour	flower colour	purple dotted	white	total
purple dotted	purple dotted	134	43	177

theor. 3 : 1   132.75 44.25

m 5.78   D/m 0.216

TABLE 29.   F$_2$ white ×

No. 1927	branch	leaf axil colour	flower colour	purple	purple dotted	'purple patched'
38.5	1	purple dotted	purple	2	27	1
,,	2	,,	,,		12	
,,	3	,,	,,		29	
,,	4	,,	,,		52	2
,,	5	,,	purple dotted		35	1
,,	6	,,	,,		22	
,,	7	,,	,,		34	
38^b.1	1	purple	purple	29		
,,	2	,,	,,	30		
,,	3	purple dotted	,,		33	2
,,	4	,,	purple dotted		57	
38.6	1	purple dotted	purple		6	
,,	2	,,	,,		22	
..	3	,,	,,		54	
,,	4	purple/purple dotted	purple dotted	3	9	1
		purple	,,	52		
,,	5	purple dotted	,,		9	
,,	6	,,	,,		23	
,,	7	,,	,,		19	
38.3	1	purple	purple	35		
,,	2	,,	purple dotted	52		
,,	3	,,	,,	21		
,,	4	,,	,,	55		
,,	5	,,	,,	25		
,,	6	,,	,,	13		
,,	7	,,	,,	17		
,,	8	,,	,,	27		
,,	9	purple dotted	purple		38	
,,	10	purple/purple dotted	purple dotted		5	
		purple dotted	,,		43	
,,	11	,,	,,		12	1
,,	12	,,	,,		19	1

purple patched

white	purple dotted + 'purple patched'	total	theor. 3 : 1		m	D/m
12	30	42	31.5	10.5	2.74	0.547
3	12	15	11.25	3.75	1.73	0.434
9	29	38	28.5	9.5	2.69	0.186
17	54	71	53.25	17.75	3.67	0.204
9	36	45	33.75	11.25	3.00	0.750
8	22	30	22.5	7.5	2.35	0.213
10	34	44	33	11	2.91	0.344
12		41	30.75	10.25	2.69	0.651
12		42	31.5	10.5	2.74	0.547
12	35	47	35.25	11.75	2.96	0.084
18	57	75	56.25	18.75	3.77	0.199
3	6	9	6.75	2.25	1.22	0.614
8	22	30	22.25	7.5	2.34	0.107
20	54	74	55.5	18.5	3.67	0.409
6	13	19	14.25	4.75	1.80	0.695
16		68	51	17	3.60	0.278
3	9	12	9	3	1.50	0
8	23	31	23.25	7.75	2.39	0.105
6	19	25	18.75	6.25	2.18	0.115
10		45	33.75	11.25	2.95	0.424
15		67	50.25	16.75	3.60	0.486
6		27	20.25	6.75	2.29	0.328
20		75	56.25	18.75	3.71	0.337
7		32	24	8	2.5	0.4
4		17	12.75	4.25	1.80	0.138
7		24	18	6	2.06	0.485
10		37	27.75	9.25	2.59	0.289
11	38	49	36.75	12.25	3.08	0.406
2	5	7	5.25	1.75	1.12	0.223
14	43	57	42.75	14.25	3.28	0.076
5	13	18	13.5	4.5	1.80	0.278
8	20	28	21	7	2.24	0.446

From this it may be concluded that purple dotted differs in 1 factor from white. From the colour of the purple F_1 of the crosses of purple patched $\times$ coloured it was already deduced that purple patched possesses the factors A_r, B, A_m and A_p. As will be further corroborated the factor for purple dotted is allelomorphic with a and consequently also with A.

Of the purple dotted F_1 plants with deviating branches of the cross white $\times$ purple patched the seeds of some branches were propagated. The F_2 data occur in the tables 29 and 30.

TABLE 30. F_2 white $\times$ purple patched

No. 1927	branch	leaf axil colour	flower colour	purple	purple dotted	white	total	theor. 3 : 1		m	D/m
45.3	1	purple	purple	39		10	49	36.75	12.25	3.12	0.721
,,	2	,,	,,	45		15	60	45	15	3.35	0.
,,	3	,,	purple dotted	23		7	30	22.5	7.5	2.39	0.209
,,	4	,,	,,	19		8	27	20.25	6.75	2.18	0.573
,,	5	,,	,,	60		22	82	61.5	20.5	3.87	0.388
,,	6	,,	,,	25		10	35	26.25	8.75	2.50	0.5
,,	7	,,	,,	51		18	69	51.75	17.25	3.57	0.210
,,	8	purple dotted /purple	purple	10	21	9	40	30	10	2.78	0.360
,,	9	purple dotted	purple dotted	1	37	10	48	36	12	3.08	0.649
,,	10	,,	,,		33	9	42	31.5	10.5	2.87	0.523
,,	11	,,	,,		27	11	38	28.5	9.5	2.59	0.579
,,	12	,,	,,		24	9	33	24.75	8.25	2.45	0.307

From the above data conclusions can be drawn on the origin of the aberrant branches. From table 29 it follows that when the leaf axil colour of the branches is identical, the progenies of these branches also correspond. The purple flowered branches give the same F_2 as the branches with purple dotted flowers, if the leaf axil colour of the two corresponds. The purple branches of which the leaf axil is purple dotted are to be taken as periclinal chimeras, of which the core is purple dotted and the epidermal skin purple. The branches on the F_1 plant with a purple dotted leaf axil colour segregate into 3 purple dotted : 1 white, while the branches with a purple leaf axil colour segregate into 3 purple : 1 white. From this it appears that from the purple dotted F_1 plant branches have developed of the genetic constitution Aa. This hereditary change is due to the factor mutation of the factor (purple dotted + purple patched) $\rightarrow A$.

Table 29 also records the F_2 data of the F_1 plant 38.3, of which the lower part of branch 10 was half purple, half purple dotted and next passed into a purple dotted leaf axil. Evidently the purple part has not formed any gametes. This was, however, the case with plant 38.6 branch 4, in which purple, purple dotted and white occur in the progeny.

In table 29 'purple patched' has also been mentioned. This refers to purple dotted plants with one sometimes two purple patches. In a certain sense I might use this name for the purple dotted plants on which purple branches occur, because I regard these two as the same phenomenon. From this it may be concluded that there are cases in which in plants with normal plasm, mutation of the factor for purple dotted $\rightarrow A$ may sporadically occur.

Table 30 gives the observations on the F_1 plant 45.3, in which besides purple and purple dotted an other colour was observed, that is an almost white flower with a strong venation of the standard not previously observed. The lower part of branch 4 was purple dotted for the flower and passed into the aberrant type towards the top. Branch 7 also passed into the aberrant type at the top. From the F_2 it appears that the departure has presumably been limited to the epidermis. The leaf axil colour of the deviating parts was normal purple. From the occurrence of the observed flower colour on various branches irrespective of each other it may be concluded that a similar form may once more arise in the later cultures.

TABLE 31. F_3 from purple dotted F_2 plants from white $\times$ purple patched

year	No.	purple dotted	fine purple dotted	white	total	theor. 1 : 2 : 1			m			D/m		
1930	123	18	37	16	71	18.5	37	18.5	3.64	4.30	3.71	0.137	0	0.674
,,	125	9	14	8	31	7.75	15.50	7.75	2.35	2.92	2.40	0.532	0.513	0.104
,,	126	5	11	4	20	5	10	5	1.94	2.35	2.00	0	0.425	0.500
,,	128	30	69	33	132	33	66	33	5.05	5.87	4.97	0.594	0.511	0
,,	124	47			47									
,,	127	35			35									

In order to trace if an actual difference can be ascertained between homozygous and heterozygous purple dotted (see § 10) an F_3 was cultivated from purple dotted plants from the F_2 of a cross white × purple patched and next the various F_3's were compared. Of the F_2's 2 were pure for purple dotted, while 4 gave a segregation as has been given in the subjoined table.

I regret to say that I had no purple dotted F_2 plants at my disposal, in which a subdivision into purple dotted and fine purple dotted had already been made in the F_2.

From the above it follows that homozygous and heterozygous F_2 plants are to be distinguished. The heterozygous ones gave a 1 : 2 : 1 segregation. In the other flower colours differing from white in 1 factor there is no difference between the homozygous dominant form and the heterozygous form.

§ 13. *Crypto purple*

Already before I mentioned a new form, which was initially observed as F_1 plant of the cross white × purple patched. The new form was called crypto purple and as it were white dotted on a purple background. Sowing out the F_1 plant gave the following result.

TABLE 32. F_2 white × purple patched

year	No.	crypto purple	white	total
1929	148	172	61	233
theor. 3 : 1		174.75	58.25	
		m 6.55	D/m 0.420	

From this it appears that in the F_2 a segregation occurred into a 3 : 1 ratio, so that it may be concluded that crypto purple and white differ in 1 factor. It is very probable that this factor is allelomorphic with a and consequently with A. The factor for crypto purple has been called A_1. A_1 was originally used by WELLENSIEK for 1 of the colour factors, but subsequently abolished (the TEDINS and WELLENSIEK, 84, 1925, p. 533).

Of the crypto purple plants in the F_2 white × purple patched some 5 were propagated. The F_3 data occur in the subjoined table.

TABLE 33. F_3 from crypto purple F_2 plants of the cross white $\times$ purple patched

year	No.	crypto purple	white	total	theor. 3 : 1	m	D/m
1930	117a	45	16	61	45.75 : 15.25	3.35	0.224
„	117b	68	21	89	66.75 : 22.25	4.12	0.303
„	117c	34	11	45	33,75 : 11.25	2.91	0.086
„	117d	39	15	54	40.50 : 13.50	3.12	0.481
„	117e	18		18			

From this it follows that also in the F_3 a 3 : 1 segregation could be established, while in 1 case the culture was pure for crypto purple. Various crosses have now been made in order to trace the genetic behaviour of crypto purple. Besides an attempt will be made at cultivating the other forms with coloured flowers : crypto rose, crypto violet, etc.

§ 14. F_2 coloured $\times$ purple patched

As already previously observed all F_1 plants of the cross purple patched $\times$ coloured were purple. The F_2 of the cross purple $\times$ purple patched occurs in the following table.

TABLE 34. F_2 purple (Solo) $\times$ purple patched

year	No.	purple	purple dotted	total
1928	304	146	47	193
theor. 3 : 1		144.75	48.25	
	m 6.04		D/m 0.207	

From this appears a segregation into 3 purple : 1 purple dotted, so that we may conclude to a monofactorial difference. As both between purple and white, purple and purple dotted, and white and purple dotted a monofactorial difference exists, we may conclude to multiple allelomorphism. In accordance with this is our previous hypothesis that the factor for purple dotted is allelomorphic with A. This factor was called A_2. (It is true A_2 was originally used by WELLENSIEK for 1 of the colour factors, but it was subsequently abolished (the TEDINS and WELLENSIEK, 84, 1925, p. 533).

From the above F_2 3 purple dotted plants were propagated. The data are given in the subjoined table.

TABLE 35.   Purple dotted F_3 of the cross purple × purple patched

year	No.	purple dotted leaf axil	colourless leaf axil	total	theor. 3 : 1	m	D/m
1929	126	67	23	90	67.5 : 22.5	4.09	0.122
,,	127	135	42	177	132.75 : 44.25	5.80	0.388
,,	128	84	29	113	84.75 : 28.25	4.58	0.164

All F_3 plants were purple dotted, while it also appeared that the 3 F_3's segregated into 3 purple dotted leaf axil : 1 colourless leaf axil. Accordingly the F_2 plants were constant for purple dotted. The F_2 of the cross pinkish white × purple patched occurs in the subjoined table.

TABLE 36. F_2 pinkish white × purple patched

year	No.	purple	purple dotted	pinkish white	pinkish white dotted	total
1930	88	121	38	43	11	213
theor. 9 : 3 : 3 : 1		119.79	39.93	39.93	13.31	
m		7.27	5.72	5.64	3.55	
D/m		0.166	0.337	0.544	0.651	

The segregation corresponds to a 9 : 3 : 3 : 1 ratio. A new form occurred, which was called pinkish white dotted. Just as is the case with pinkish white, this new form is hard to distinguish from white. On observing this case we should pay special attention to the dull rose dotted leaf axil colour. The new pinkish white dotted form has the constitution $A_2 A_2 a_m a_m$.

The cross light purple × purple patched gave the following F_2 data (tab. 37).

Though for a similar polyhybrid segregation the figures ought to be supplemented, yet we may conclude to a 27 : 9 : 9 : 9 : 3 : 3 : 3 : 1 ratio. As new forms violet dotted ($A_2 A_2 a_r a_r BB A_m A_m A_p A_p$)

and light purple dotted ($A_2\ A_2\ a_r\ a_r\ bb\ A_m\ A_m\ A_p\ A_p$) occurred.

TABLE 37. F_2 light purple × purple patched

year	No.	purple	vio-let	rose	purple dotted	light purple	violet dotted	rose dotted	light purple dotted	total
1930	77	65	20	24	24	10	8	9	1	161
,,	81	84	30	29	25	9	11	12	3	203
total		149	50	53	49	19	19	21	4	364
theor.		153.63	51.21	51.21	51.21	17.07	17.07	17.07	5.69	
m		9.52	6.64	6.61	6.66	4.02	4.02	4.01	2.37	
D/m		0.486	0.182	0.271	0.332	0.480	0.480	0.980	0.713	

To the above observations may be added the F_2 of apple × purple dotted.

TABLE 38. F_2 apple × purple dotted

year	No.	purple	apple	purple dotted	apple dotted	total
1930	85	85	30	27	12	154
,,	86	104	33	35	13	185
total		189	63	62	25	339
theor.		190.8	63.6	63.6	21.2	
m		11.52	7.19	7.21	4.43	
D/m		0.106	0.083	0.222	0.858	

From the table follows a dihybrid segregation.

Though a number of observations should be added to the above; yet from the data a hypothesis on the heredity of the mosaic type may be set up in outline. This will be found in the next §.

§ 15.   *The genetic constitution of the purple patched plant*

From the above the conclusion was drawn that the purple patched plants consist of a mosaic of genetically different parts. In a purple patched plant may occur purple patches on a white background and besides parts may be purple dotted. It was ascertained that in the purple parts the factor A occurs, that in the dotted parts (both in the coloured and the colourless cells of them) the factor A_2 occurs.

In the reciprocal crosses of white (aa) × patched it was shown that the patched type only occurs, if patched is used as mother plant, whereas in the case that in this cross patched is used as father plant all coloured descendants are purple dotted. From this it was inferred that the factor for purple patched is the same as the one for purple dotted, that is A_2. The difference in colour, however, is due to difference in the plasmatic condition. To interpret the hereditary phenomena I have been obliged to assume a labile factor A_2 and labile plasm. In connection with this I always spoke of the factor for (purple dotted + purple patched), while the observations in the tables are joined into one group.

The colourless cells of the patched form have the constitution $A_2 A_2$, the plasm of these cells being labile. The purple cells have the constitution $A A_2$, and the purple dotted parts the constitution $A_2 A_2$, the plasm being normal. From this hypothesis it follows that if the plasm of an $A_2 A_2$ plant continues labile and no change occurs in the factor A_2, the flowers will be colourless, that is white. Such plants have repeatedly been observed; they could, however, be distinguished from white aa forms by their patched leaf axils. Most plants with labile plasm showed purple patches as a result of the factor mutation $A_2 \rightarrow A$. Moreover dotted parts appeared as a result of the transition of the labile plasm to normal plasm.

Purple dotted and purple patched are fundamentally different. Purple dotted is a pattern of genetically equal parts, purple patched on the other hand consists of a mosaic of genetically different parts. The labile plasm furthers the mutation of $A_2 \rightarrow A$. From the occurrence of forms patched to a varying degree, it may be concluded that the frequency of the factor mutation is affected.

The possibilities which can be realized in the same plant, occur in the subjoined table.

TABLE 39.   Scheme of the mosaic of the patched plants

zygote	plant		
	genes	plasm	flower
$A_2 A_2$ + labile plasm	$A_2 A_2$	labile	colourless
	$A_2 A_2$	normal	purple dotted
	$A A_2$	labile	purple
	$A A_2$	normal	purple

With the above hypothesis all observations can be made to fit in, except one observation which renders the formation of an auxilliary hypothesis necessary. It concerns the white dotted purple patches previously discussed, appearing in flowers which are patched as well as purple dotted.

According to the above scheme the colourless cells have the constitution $A_2 A_2$ + labile plasm. From the shape of the patch it may be inferred that it arose from 1 cell. From this it may be concluded that all cells will possess the constitution for purple $A A_2$. Therefore I cannot explain the above in any other way than by assuming that the colourless cells are due to back mutation from $A A_2$ to $A_2 A_2$.

The hypothesis set up explains how different flower colours can occur in one plant, viz. purple patched on white background, purple dotted and almost white. Further it explains how in the progeny of purple patched by the side of purple patched, purple, purple dotted and almost white can occur in a varying percentage and also the difference in behaviour of the reciprocal crosses with purple patched, from which it appeared that the purple patched type was not inherited by the father, that, however, all F_1 plants with purple patched as father were either purple or purple dotted and consequently in the normal plasm purple dotted appeared instead of purple patched. Besides the hypothesis explains that constant purple and constant purple dotted forms have been cultivated from the progeny of the purple patched forms. In the discussion (sektion 4) various points will be discussed more fully.

§ 16. *Historical survey of the patched and dotted flower colours*

As previously mentioned the purple patched form arose but once in a large culture of purple-flowering Fletumer rozijnerwten.

From the research described here it appeared that the purple patched form, however, cannot be taken as a simple factor mutant, but that it differs from the ancestral form in 2 points. With respect to the origin of the purple patched form it must be assumed that the ground factor for colour A mutated to A_2 and that moreover the normal protoplasm passed into the labile state.

Most likely the two hereditary changes occurred in 1923, purple dotted ($A_2 A_2$) likewise not being observed before.

The other flower colours are du e to factor combination. The formulas of the new flower colours mentioned in this chapter occur in the subjoined table.

TABLE 40.   Formulas of the patched and dotted flower colours

purple patched	$A_2 A_2 A_r A_r B B A_p A_p A_m A_m$ (labile plasm) (pl. II, fig. 8)
rose patched	$A_2 A_2 A_r A_r b b A_p A_p A_m A_m$ (labile plasm)
purple dotted	$A_2 A_2 A_r A_r B B A_p A_p A_m A_m$ (normal plasm) (pl. II, fig. 12)
violet dotted	$A_2 A_2 a_r a_r B B A_p A_p A_m A_m$,,
rose dotted	$A_2 A_2 A_r A_r b b A_p A_p A_m A_m$,, (pl. II, fig. 8)
light purple dotted	$A_2 A_2 a_r a_r b b A_p A_p A_m A_m$,,
apple dotted	$A_2 A_2 A_r A_r B B a_p a_p A_m A_m$,, (pl. II, fig. 7)
pinkish white dotted	$A_2 A_2 A_r A_r B B A_p A_p a_m a_m$,,
crypto purple	$A_1 A_1 A_r A_r B B A_p A_p A_m A_m$,, (pl. II, fig. 11)

In the same way as in table 10 the patched and dotted flower colours mentioned in this chapter have been mentioned in the subjoined table.

TABLE 41.   Survey of the occurrence of the various flower colours

factormutation + plasm mutation	factor combination	
purple → purple patched 1923	rose patched	1930
	purple dotted	1924
	crypto purple	1928
	apple dotted	1929
	pinkish white dotted	1929
	violet dotted	1929
	rose dotted	1929
	light purple dotted	1929

In the further research various other patched and crypto forms will most probably arise through factor combination.

In literature a short report of KAZNOWSKY (53, 1926, p. 1) of a form with white spotted flowers occurs. Of the genotype of this form nothing is known to me.

§ 17.   *The seed colour and the seed form of the purple patched plants*

Besides in the flower colour and in the leaf axil colour the factor for patching is also manifested in the colour of the seedcoat. The seeds have a very marked patching pattern. On the purple branches of purple patched plants they are normal marbled and of a perfectly identical pattern with the Fletumer rozijnerwt, from which purple patched arose. The seed colour of the purple dotted plant deviates greatly from the marbled type. The seeds are dotted especially so round the hilum.

The seed form is affected by the ground factor for flower colour. The seeds on the purple branches are always angular, while the purple dotted forms bear round seeds, just like all purple dotted plants.

Sektion 4. Discussion

First of all I wish to discuss more fully 2 publications of great importance for the problem of the mosaic plants. CHITTENDEN's publication (13, 1927, p. 416) is interesting, because he tried to explain the origin of the patches in quite a different way from what had been hitherto adopted, while the EYSTER genomeric theory (32, 1928, p. 670) attempts to give an explanation of the mechanism of factor mutation.

The striking resemblance the behaviour of the mosaic flowered plants shows to that of the non-mendelian variegations on the one hand, and the researches by GUILLERMOND in which it was determined that the anthocyanin is formed in plastids, on the other hand, justify CHITTENDEN's investigation into the literature (13, 1927, p. 415) in order to trace whether the occurrence of the mosaic flower colour might be based on plastid inheritance. CHITTENDEN was actuated to this by his objections to the assumption of labile factors and because he disapproves of endeavouring to force facts into line with a theory by making such assumptions as the lability of the gene. Of late years, however, so numberless cases of labile genes have been described, that objections to them need not be further refuted.

He premises that the forms used for crossing are not 'crypto mosaic'. Unlike CHITTENDEN I am of opinion that nearly all cases from literature meet this just requirement, that the unicoloured forms used for crossing, originate from pure lines and that they have no history of mosaicism in their ancestry and when crossed reciprocally with similarly pure white and self coloured varieties no mosaics should appear in their descendants.

CHITTENDEN's opinion is that many of the published cases may be more easily interpreted on the basis of a plastid inheritance. He draws his data partly from MARRYAT's researches (58, 1909, p. 32).

MARRYAT did not succeed in placing her observations on the flower mosaic of *Mirabilis* on a factorial base. This must not be considered an argument in support of non-mendelian inheritance, as it has appeared from CORRENS' researches that there does exist a factor for patched flowers in *Mirabilis*.

CHITTENDEN regards the patched flowered plants on the analogy of variegation as plants in which the genetical constitution of the nucleus is white, the colour being due to the fact that the cells of the coloured areas contain autonomous anthocyanin-producing plastids, whereas the plastids of the white areas are distinct and are normally subject to the nucleus. He regards this conception as highly suggestive; I cannot subscribe to this view. He drew a wrong conclusion, as from the appearance of hereditary variegation has been inferred that in the nucleus factors are localized for the formation of chlorophyll and the non-mendelian variegation is brought about by the fact that the plastids are not capable of forming chlorophyll (are 'ill'). For the formation of chlorophyll it is required that the hereditary factors for its formation are present and besides that the plastids are capable of forming chlorophyll. On the analogy of this the assumption that a plant that does not possess the factors for pigment, is yet capable of forming anthocyanin, is certainly unacceptable.

CHITTENDEN points out that of various plants of which non-mendelian variegation is known, the flower mosaic would be explained by a mendelian factor. He doubts these views, because almost all the crosses of which the results are published are those in which the mosaic form was the female parent. This observation is right; in the uncommonly extensive research for instance of BAUR on *Antirrhinum* only the outline was given and it cannot be deduced from his communications on the flower mosaic whether BAUR also made reciprocal crossings. I do not doubt this but like CHITTENDEN I have felt the suppression of the figures found in the analyses as a serious objection in studying the data. Some reciprocal crosses, however, are known from WHELDALE's research, from which it has appeared that it does not matter whether the patched *Antirrhinum* is used as mother or as father.

An argument against CHITTENDEN's assumption certainly is, that patched flowers occur with varying backgrounds, for instance in *Mirabilis* :

pale yellow   patched with yellow.

magenta rose   „   „   orange red.

white   „   „   yellow or magenta.

These cases cannot be explained by CHITTENDEN's hypothesis, while they are plausible when one of the colour factors is assumed to be labile.

In the older literature cited by CHITTENDEN the occurrence of periclinal chimeras was not known, and if, therefore, in an exceptional case as in WHELDALE's researches, in which an aberrant 'yellow striped with pale crimson plant' was used as father, no patched form appeared in the progeny, it is quite possible that we had to deal with a case in which the epidermis was patched and the subepidermis was not. When the same plant has not likewise been used as mother, no conclusions must be drawn from it.

From the above it may appear that in my opinion CHITTENDEN has deserved well of us for directing the attention to the resemblance between mosaic and variegation, but that his view that most of the phenomena recorded in connection with mosaic may be interpreted on the basis of plastid inheritance is not supported by literary data.

An other theory interesting for the study of flower mosaic is EYSTER's.

EYSTER (32, 1928, p. 670) agrees with many other investigators that the flower mosaic is based upon genetic differences of the same factor, but they differ in their opinion on the mechanism that leads to the origin of these genetic differences. EYSTER assumes somatic segregation of heterogenous genomeres occurring in one gene, others on the other hand, speak of somatic mutation. EYSTER regards the genetic changes of the factor for mosaic as a result of dissimilar genomeric segregation not as mutations, but rather as the expressions of genes with the contrasting genomeres in different numerical combinations, quite like the origin of individuals with new combinations of mendelian characters due to recombinations of the genes or entire chromosomes. If the word mutation is used in a generic sense to mean any hereditable change, the hereditary changes of the factor for mosaic assumed by EYSTER come under the head mutation, so that I regard the EYSTER theory as an attempt at giving an explanation of the ever-mutating factors, which in many instances underlie the flower mosaic.

EYSTER, however, objects to the adoption of ever-mutating factors because in his opinion mutations are too rare and erratic to account for the precision which accompanies the origin and genetic relationships of the types which make up a mosaic series. In his opinion a factor for mosaic is due to mutation of one or more of the genomeres and genetic differences of the flower mosaic are owing to the distribution of these deviating genomeres.

To EYSTER's views I must raise several objections, which I shall discuss below and illustrate by as many examples as possible.

In the case of *Antirrhinum* analysed by BAUR the various types of flower mosaic joined to pal_{rec} would have the following constitution in EYSTER's symbolization:

$$\begin{array}{lll}
Pal\ldots\text{k}\quad\text{C} & & \text{unicoloured red.} \\
& \left\{\begin{array}{l}(\text{k---m 1})\ \text{C} + \text{m1c} \\ (\text{k---m 2})\ \text{C} + \text{m2c} \\ (\text{k---m 3})\ \text{C} + \text{m3c}\end{array}\right. & \begin{array}{l}\text{large red patches.} \\ \text{medium large patches.} \\ \text{small red patches.}\end{array} \\
pal\ldots\text{k}\quad\text{c} & & \text{elfenbein.}
\end{array}$$

To this *Pal*-series, however, belong various other multiple allelomorphs, which cannot be designated with EYSTER's symbolization. This can only be done by assuming that the number of genomeres of which a factor is constituted, is not constant, but that in cases of stabile multiple allelomorphs each of these genes possesses a constant number of genomeres, and consequently but one form, which is recessive with respect to all others, can possess the same number of one of these genes.

There is an inconsistency between the genomeric constitutions and the dominance relationships derived from them. EYSTER sets up series beginning with kC as stabile and dominant and finishing with kc as stabile recessive. Between these 2 cases there lies a series of genomeric combinations, of which EYSTER determined that the genes with a smaller number of C genomeres dominate over those with a larger number of C genomeres. This is incompatible with the fact that kC in which not a single c occurs, should be dominant and kc in which only c genomeres occur is recessive to all others. From this it follows that the arrangement of the multiple allelomorphs by EYSTER departs from the current order, as it is usual to arrange the multiple allelomorphs in a series in such a way that a preceding one dominates over a succeeding one. From this it would follow that in

EYSTER's series kC should be followed by (k-nn-mm) C + (nn + mm)c and would constantly decrease in the number of c genomeres.

If it is correct that the genes do not consist of a complex unity, but of a number of genomeres constant for each factor, it must be accepted that during the origin of factor mutants all genomeres mutate, so that in consequence of the simultaneous mutation of nn + mm genomeres the factor kC passes into kc. For it is not likely that the factor mutants pass through a labile stage and next after somatic genomeric segregation pass into stabile factors. If this were correct, the number of labile factors would be ever so much larger.

It might be that only the genes for mosaic were built up of genomeres, but in that case the origin of *alba* forms could not be explained, since pure white has never been stated in the progeny of mosaic plants. The white plants in the progeny of mosaic always appeared to be crypto mosaic in the sense CHITTENDEN takes it. The occurrence of pure white might be expected according to EYSTER's views, as he often observed white in the somatic cells and a similar constitution might therefore be expected in part of the gametes.

In the above I advanced various points against EYSTER's views. From those it appeared that the explanation of the ever-mutating factors by somatic genomeric segregation presents difficulties.

From the discussion of the patched pea it has appeared that it consists of a mosaic of genetically differing parts. These hereditary differences in one and the same plant were explained by the assumption of somatic factor mutation. From this it follows that the hereditary behaviour of the patched flowers in the pea is fundamentally different from the cases in which the patching-pattern is based upon the differentiation of a pattern, of which therefore the coloured as well as the colourless parts are genetically identical and which is inherited by a stabile hereditary factor. It was demonstrated that the purple patches were heterozygous for the ground factor for colour, while in the colourless cells this factor did not occur.

Corresponding explanations have been given by BAUR (5, 1914, p. 205) for the case of the patched *Antirrhinum*s and by DEMEREC (30, 1926, p. 25) for the patched *Delphinium*s. While in *Antirrhinum* the patched flowers are bicoloured as a result of one ever-mutating factor and it is very likely that the tricoloured *Mirabilis* flowers are due to the mutation of 2 factors, this argument must not be simply adopted for the explanation of the flower mosaic of *Verbena* examined by EYSTER by assuming more mutating factors in that case. From EYSTER's data it may be concluded that in *Verbena* the various colours in one plant can be explained in the same way as was done for *Antirrhinum*, *Delphinium* and *Pisum*, with this difference that one factor can mutate to various other factors.

In the genetic behaviour of the patched flowered peas I could not find any argument which might support the genomeric theory. Indeed in the selfings of patched plants a number of white flowered plants were observed, but it could always be ascertained from the patched leaf axils, that the plant was patched, while in the progeny of these 'white' forms there never arose pure white (aa) forms, as might be expected according to EYSTER's theory. In my opinion the genomeric theory should not be applied to the case of the patched *Pisum* and the factor mutation should be regarded in the same way as BAUR and DEMEREC did.

The direction of the mutation was always from the recessive A_2 factor to the dominant A. From the appearance of purple branches and larger purple complexes in the strongly mosaic plants and the non-appearance in slightly mosaic plants it may be concluded that the point of time of mutation in the development of the different parts of the plant presumably does not vary, while from the frequent occurrence of many small spots in the strongly patched forms it may be concluded that the frequency of mutation during the various stages is about equal.

It was never observed that both the A_2 factors mutated; whether this is chance or whether mutation of one factor furthers the stability of the other, I must leave undecided.

Neither have any cases based upon plastid inheritance been described in the literature that appeared after CHITTENDEN's publication. CAYLEY (10, 1928, p. 529) indeed, communicated that the flames of 'broken' tulips can be transmitted by inoculation. In how far reciprocal crosses are different, I do not know.

The only case that might possibly supply arguments in support of plastid-inheritance is the patched pea, because it has appeared that the mosaic character of the flowers is not inherited with the father.

If the mosaic in the pea was really founded upon such a base, as CHITTENDEN assumed, no patched plants might appear in the progeny of purple branches (with purple leaf axils) of a purple patched plant or in the progeny of a purple plant with purple leaf axil arisen from the self-pollination of patched. This, however, is the case; a monofactorial segregation does occur. From this behaviour the conclusion was drawn that purple patched in the pea was due to a labile factor and not to plastid inheritance.

Above the appearance of purple dotted plants in the progeny of purple patched peas was mentioned. It is very interesting that in BAUR's cultures of mosaic *Antirrhinum*s (7, 1930, p. 166) forms arose which were called *maculosa* and the pattern of which reminds us strongly of the purple dotted pattern. *Maculosa* appeared to be a constant form, while the factor for *maculosa* appeared to be multiple allelomorphic with the labile factor for patched and the dominating factor *Pal* which often arises from it through mutation.

In discussing the relation between purple patched and purple dotted the hypothesis was set up that the labile state of the plasm was conductive to the mutation of A_2 to A. Further it was accepted that the plasm was not transmitted by the father. Part of the purple dotted F_1 plants of white $\times$ purple patched, however, had several purple branches. From this it follows that A_2 can also mutate in the heterozygous $A_2\,a$ plant with normal plasm. In how far this is due to the influence of the factor a, or whether it must be explained in an other way, is still being investigated. In the F_2 of the above mentioned purple dotted F_1 plants, I have but rarely found aberrant branches. The circumstances were, however, much more unfavourable, because the F_1 plants were strongly branched and the F_2 plants owing to crowding formed but 1 or 2 lateral branches. In the same F_1 plants an occasional purple patch was observed on the flower, from which it appears that not only in the branching but also during the flower development mutation may sporadically occur.

For the rest the progeny of such purple branches behaved normally, the purple branches gave a 3 : 1 segregation, patched plants never having arisen from them. From this it appears that not only the labile plasm of the patched plant is conducive to mutation, but that also in A_2a plants with normal plasm mutation sporadically occurs. In how far this is due to the influence of the recessive a factor is being investigated.

In the F_2's of crosses with purple patched as mother it could be ascertained that purple patched plants occurred but in a very small percentage. Nearly all A_2A_2 plants were dotted. From this it may be inferred that after crossings in nearly all cases the plasm passes to the normal state.

Various questions are still unanswered in the study of the mosaic pea; possibly part of them can be answered in the continued research.

CHAPTER III

LENGTH FACTORS

§ 1. *Introduction*

Most researches on the heredity of the length of stem refer to the genotypical difference between one of the many short forms and one of the many tall forms. In these crosses it has been established that tall dominates and that the difference in length is determined by the growth favouring factor *Le*, corresponding with the monofactorial difference shown by MENDEL (59, 1866, p. 14). Genetical analysis of mutually differing tall forms, however, have rarely been made.

MENDEL (59, 1866, p. 14) pointed out that the F_1 plants usually exceed the parents in length. He mentioned crosses of one foot and 6 feet tall plants, the F_1 of which oscillated between 6 and $7\frac{1}{2}$ feet.

DARWIN (28, 1876, p. 163) mentioned various crosses of short forms. These crosses were made to trace in how far the F_1 type differed from that of the parents. In many cases he found that the F_1 type corresponded with one of the parents, in an exceptional case the F_1 surpassed the taller of the parent plants.

KEEBLE and PELLEW (54, 1910, p. 49) studied the cross Autocrat (thick stem, short internodes) × Bountiful (thin stem, long internodes), which yielded an F_1 generation considerably taller than either parent grown under like conditions. The F_2 could be subdivided into 4 groups, viz. the F_1, the Autocrat, the Bountiful and the Dwarf type, while a $9 : 3 : 3 : 1$ ratio could be ascertained. They explained this segregation by assuming that 2 factors were involved, 1 factor for length of internode, the other for thickness of stem.

NILSSON—EHLE (63, 1911, p. 8) directed the attention to the fact that most probably by the side of the easily identified factor *Le*, there also occur such as cause smaller hereditary differences, so that by the side of the distinct 3 : 1 segregation a subdivision occurs, both in the tall and in the short forms.

WHITE (95, 1917, p. 546) pointed out that many dwarf varieties have a thick robust stem, though they have comparatively short internodes and only few in number. He regards the factor T of KEEBLE and PELLEW as not for thickness of stem, but as a factor for large number of internodes. WHITE distinguishes the following types:

tall, large number of long internodes	$TT\ Le\ Le.$
half dwarf, large number of short internodes	$TT\ le\ \ le.$
half dwarf, small number of long internodes	$tt\ \ Le\ Le.$
dwarf, small number of short internodes	$tt\ \ le\ \ le.$

In a subsequent communication (96, 1917, p. 174) WHITE pointed out the danger of using different varieties, because drawing conclusions on the transmission of differences in length is open to criticism from the hypothesis of multiple factors. The same phenotype that is determined by different factors can bring about considerable confusion in an attempt to determine the factorial groups of *Pisum*.

The TEDINS (82, 1923, p. 355) crossed Ragunda × 0234 and inferred from the transgressions in the F_2 that it is evident that more factors concerning the length of internodes must be involved in the crosses.

KAPPERT (52, 1929, p. 84) surmises that the number of internodes is determined by a number of quantitative factors, the action of which can be transgressive.

A number of investigators ascertained an excess of short plants in the F_2. To what this is due, is not yet known.

Of late more attention has been paid to the reciprocal crosses of short plants. As the length of the stem of the peas is important for practical purposes, it is desirable that the known forms are studied as systematically as possible. It has appeared that growth inhibiting factors play a prominent part. At about the same time 2 cases of growth inhibiting factors were described. The one refers to the crypto dwarfs which appeared in a cross of two short forms (RASMUSSON, 67, 1927, p. 40), the other to the slender peas which occurred in my cultures in a cross of two short forms (39, 1927, p. 481). The two forms are entirely different, as will be further discussed.

From the above it may be inferred that there exist more factors for length. The data of MENDEL, DARWIN, KEEBLE and PELLEW, and of WHITE point out that besides the growth favouring *Le* factor there are one or more additional factors likewise exercising a growth favouring influence; the researches of RASMUSSON and myself proved that there also occur growth inhibiting factors. This, therefore, confirms the hypotheses of NILSON-EHLE, of the TEDINS and of WHITE that more than one factors determine the length.

§ 2. *The polymeric, growth inhibiting factors La and Lb*

In a preliminary communication I drew the attention to a case of polymeric (multiple) growth inhibiting factors (39, 1927, p. 481). This was inferred from the occurrence of an extremely tall, double recessive form in the crosses of short forms. This new form was called slender pea. The slender pea first appeared in my cultures in 1926. In the same year various crossings were made, the F_1's of which were cultivated in 1927, a great many of their seeds being sown out in autumn in order to enable us to determine the segregations in the germ plants the same year. Because the germ plants require very little room and were sown early in September on beds from which an other culture had been cleared away, it was possible to grow large cultures. Every year some new crossings are made and other forms of various origin added to the material in order to obtain the completest possible survey of the distribution of the factors of length. At the same time I got an opportunity of further examining various questions which had risen with reference to the previous publication.

In 1926 the original crossings were repeated in order to get more data on the numerical ratio of the short and slender peas in the F_2 of the cross short 202.1 with short 201.1. Here again the F_1 type much resembled the 201.1 one and accordingly it was short just as the two parents. The 6 F_2's have been mentioned in table 42, the totals of the 7 F_2's from the previous communication being added to it. From this it appears that from a total of 2884 F_2 plants 2705 belonged to the short type and 179 to the slender type. In a 15 : 1 ratio it might be expected that theoretically 2703.8 were short and 180.2 slender. The figures observed differ extremely little from those theoretically expected. A germinating dish in the subjoined illustration plainly shows the great differences in the germ plant types of the above F_2. In this

TABLE 42.   Segregation into short and slender of the cross
202.1 × 201.1

year	No.	short	slender	total
1927	17—23	552	40	592
1928	243	198	10	208
,,	253	530	30	560
,,	254	387	37	424
,,	255	446	32	478
,,	258	205	10	215
,,	262	387	20	407
total		2705	179	2884

theor. 15 : 1   2703.8   180.2

m   13   D/m   0.092

dish there occurred 132 short plants and 9 slender plants, corresponding with the expected 15 : 1 ratio.

From the 15 : 1 ratio I inferred (1927, p. 481) that the slender peas
are double recessive for 2 factors, of which either of the parents
possesses one in a dominant condition. The factors for slender were called la and lb. The slender pea got the formula $la\ la\ lb\ lb$; 201.1 the formula $la\ la\ Lb\ Lb$ and 202.1 the formula $La\ La\ lb\ lb$.

The data on the cross of the slender pea with the parents were supple-mented. Table 43 gives a survey of

Fig. 3.   F$_2$ of short 202.1 × short 201.1.

the figures referring to the cross of 201.1 and slender. From the figures it appears that they come up to the theoretically expected 3 : 1
ratio.

TABLE 43. Segregation into short and slender of the cross
201.1 × slender

year	No.	short	slender	total
1927	48.1	115	36	151
1928	48.2	97	32	129
,,	48.3	183	69	252
,,	97.1	265	84	349
,,	97.3	243	82	325
total		903	303	1206
theor. 3 : 1		904.5	301.5	

m   15.02   D/m   0.100

Table 44 contains the supplementary data on the cross 202.1 and slender. These figures also correspond with the theoretically expected 3 : 1 ratio.

TABLE 44. Segregation into short and slender of the cross
202.1 × slender

year	No.	short	slender	total
1927	36.1	136	39	175
1928	84.1	234	81	315
,,	84.2	194	71	265
total		564	191	755
theor. 3 : 1		566.3	188.7	

m   11.87   D/m   0.194

To give a representation of the various phenotypes I subjoin the measurements of the internodal lengths of the main stems of the short forms 201.1, 202.1, and 203.1, further of the slender form and of the short F_1 201.1 × 204.1.

TABLE 45. Internodal length of 201.1 (*la la Lb Lb*)

year	No.		total
1930	12.1	1 1 1 2 2 2½ 3½ 3½ 5 5 5½ 5½ 5 6 5½ 6 6 5½ 4½ 4 3½ 3 2 2	90½ cm
,,	12.2	1 1½ 1½ 1 2½ 3 3 4 5 5 6 5 5 6½ 5½ 5½ 6 7 6 5 2 2½ 3	92½ ,,

TABLE 46.   Internodal length of 202.1 (*La La lb lb*)

year	No.		total
1930	11.1	1 ½ 1 1 1½ 1 2 3 2½ 3 3 2 3½ 3½ 3 4 4 4½ 4 4 5 4½ 4 3½ 3	72 cm
,,	11.2	½ 1½ 1 1 1 1 1 1 1½ 1½ 2 2 3 3½ 3 3 4 5 4½ 4½ 5 5½ 5 5 4 3½	72½ ,,

TABLE 47.   Internodal length of the slender pea (*la la lb lb*)

year	No.		
1929	23	5 8 8½ 8½ 11 9½ 10 14 13½ 15 13½ 15 13½ 14 13 13½ 13 11½ 11 11½ 10½ 12 10 11½ 9½ 10 8½ 7½ 6½ 7½ 6 6 6 5½ 5½ 5½ 3	total 363½ cm
1929	20	3½ 6 3½ 5 5 5½ 5½ 10 8 9 9½ 10 11½ 12 14 15 13½ 14½ 13 12 13 11½ 13½ 13 10½ 11 11½ 9½ 8 7½ 9 7 3½ 3	total 317½ cm

TABLE 48.   Internodal length of 204.1 (*La La lb lb*)

year	No.		total
1930	16.1	1 1 1½ 2 2 3 3 4 3½ 3½ 3 4½ 5 4½ 5 5 5 4½ 4½ 4 3½	73 cm
,,	16.2	1½ 1 1 1½ 2 2½ 3 2½ 4 3½ 4 5½ 5½ 5 4 4 4½ 4 4 3½ 3 3	72½ ,,

TABLE 49.   Internodal length of the F₁ 201.1 × 204.1 (*La la Lb lb*)

year	No.		total
1930	39.1	1½ 1 1½ 1½ 2 2 3 2½ 3 4 4½ 5 6½ 6½ 6½ 5½ 6 5½ 6 5 5 4½ 2	90½ cm
,,	39.2	1 1 1½ 2 2½ 2½ 4 3½ 3½ 5 6½ 7 7 6 6½ 7 6½ 6 6 5½ 3	93½ ,,
,,	39.3	1 1 1½ 1½ 2 2½ 3 3 3½ 4½ 7 6 6½ 6½ 7 6½ 7 6 5½ 3½ 3	88 ,,
,,	39.4	1½ 1 1 1½ 2 2 2½ 2½ 3 3 4 3½ 7 5½ 6 5½ 6½ 5½ 6 6 6 5½	87 ,,
,,	39.5	1 1 1½ 2 2½ 3 2½ 4 5 8 7 5 6½ 6½ 6 7 6 6½ 6 6 3	96 ,,
,,	39.6	1 1 1 1½ 2 2 3 3 3 4 3½ 5 6 6½ 5½ 5½ 5½ 5½ 5½ 5 5 5 4 2	91 ,,

The average length of stem of the slender peas (*la la lb lb*) may be put at ± 335 cm, of 201.1 (*la la Lb Lb*) at ± 90 cm, both of 202.1 and 204.1 (*La La lb lb*) at ± 70 cm, and of the F₁ 201.1 × 204.1 (*La la Lb lb*) at ± 90 cm. From this it follows that the factor *Lb* has an inhibiting effect on the growth and reduces the length of the stem from 335 cm to $\pm$ 90 cm, while the factor La has a still more inhibitory effect on the growth and reduces the length of the stem from 335 cm to $\pm$ 70 cm. The F_1 on the other hand, on which both Lb and La have an inhibitory effect is little distinguished from the 201.1 type. On the whole it may be inferred from the data that the internodes in the flower bearing part of the stem of the heterozygote are longer than in 201.1. From this it might be inferred that the factors La and Lb reduce the inhibitory effect on the growth of $La\ La$ or $Lb\ Lb$. It is, however, very probable, that more factors are involved. The fact for instance that different $La\ La\ Lb\ Lb$ short forms (groene erwt, pois à brochettes, recessief geel, paarse capucijner, stamcapucijner, etc.) all differ in length, points to the fact that there exist still more growth modifying factors. In order to enable us to draw decisive conclusions concerning the interaction of La and Lb, it should be ascertained that the forms $la\ la\ Lb\ Lb$ and $La\ La\ lb\ lb$ used for crossing are homozygous and identical for the other factors. For the present the analysis is not yet far enough advanced to enable us to do so.

In genetic literature various cases have been explained by assuming growth inhibiting factors. DAVENPORT (29, 1917, p. 340) demonstrated for the body length in man certain dominant growth repressing factors. TAMMES (80, 1911, p. 242) investigated the transmission of the seed length in $Linum$, determined some polymeric factors for it and pointed out the importance of growth inhibiting and growth favouring factors.

There is a vast literature on the polymeric factors, but it presents various problems which should be further investigated. The polymeric factors are in some cases equivalent in other cases non equivalent; they can have a cumulative effect or not.

SHULL (72, 1914, p. 137) connects with the case of the capsule form in $Capsella\ Bursa\,pastoris$ views on the presumable origin of the two polymeric factors. He ascertained that the two factors are completely equivalent and must presumably be taken as identical factors which were originally allelomorph, but which were localized through transfer to non homologous chromosomes.

SIRKS (70, 1929, p. 806) thinks the explanation of various cases with the aid of polymeric factors rather doubtful. He bases his objections on considerable departures of the figures observed from the theoretically expected segregation figures, nothing being known of the localization. In my opinion, therefore, it is important to trace in how far the segregation figures found do agree with the theoretically expected figures and what arguments may be brought forward for a localization in different chromosomes.

As to the above-mentioned 15 : 1 ratio it may be deduced from the F_2 figures, that the departures lie within the limits of error, while from the F_2 of tall and slender a corroboration of the 15 : 1 ratio was obtained. From the 15 : 1 ratio it was deduced that the slender type is determined by 2 independent factors, which are localized in different chromosomes. That this is most probably a case of independent transmission of the two factors La and Lb is likewise confirmed by a case to be discussed afterwards, of linkage of a wax factor with one of the factors for slender and an independent transmission with respect to the other factor for slender.

From the various crosses of short types it has appeared that the remaining not yet analysed hereditary factors for length have little effect in comparison with the striking growth inhibiting effect of the La and Lb factors. The growth inhibiting factors found, the double, recessive combination of which was obtained from crosses, however indicate that we must not conclude to the effect of the other not yet sufficiently analysed factors without reserve. The interaction of the factors for length should be separately determined for each case, and therefor it is necessary to make as many new combinations as possible. With a view to the length research it is important to have at our disposal genotypes which are recessive for the greatest possible number of factors.

§ 3. *Interaction of the length factors Le, La and Lb*

In the above 2 growth inhibiting factors were discussed. While the *le le la la lb lb* forms, in which the growth favouring factor *Le* did not occur, attained a length of 3.35 m. and more, a single one even a length of 4.30 m., it was important to trace what phenotype the plants showed of the genotypical constitution *Le Le la la lb lb*, in which, therefore, *Le* occurred as growth favouring factor.

For this purpose I crossed a tall pea (205.1) with short 201.1 and short 202.1. The F_1 plants belonged to the tall type. In the F_2 of tall and short 201.1 and tall and short 202.1 no slender plants appeared, so that it was concluded from this that the tall form possessed besides Le also the factors La and Lb.

The F_1 of the cross tall × slender belonged to the tall type. As tall possesses the 3 length factors Le, La and Lb and the slender pea is recessive for these 3 factors, there existed a trifactorial difference. The subjoined table gives the figures observed. In a total of 4228

TABLE 50. Segregation into tall, short and slender of the cross tall × slender

year	No.	tall	short	slender	total
1927	34.1—47.1	410	152	33	595
1928	233	385	142	32	559
,,	245	500	190	52	742
,,	249	181	84	15	280
,,	257	287	102	34	423
,,	265	289	77	33	399
,,	272	283	74	25	382
,,	275	190	48	14	252
,,	284	432	129	35	596
total		2957	998	273	4228
theor. 45 : 15 : 4		2972.7	990.9	264.2	
m		29.66	27.33	4.13	
D/m		0.529	0.261	2.131	

F_2 plants 2957 tall plants were stated, 998 short and 273 slender, the theoretical ratio for that number being 2972.7 : 990.9 : 264.2. The departures lie within the limits of error, so that we may infer an independent transmission. Besides it may be concluded from the 45 : 15 : 4 ratio, that the $Le\ Le\ la\ la\ lb\ lb$ forms belong to the slender type. Here, therefore, we have to deal with the case that in the $Le\ Le\ la\ la\ lb\ lb$ genotype tall does not dominate; on the other hand tall does dominate if one of the factors La or Lb is present. Though corresponding in length with slender, the $Le\ Le\ la\ la\ lb\ lb$ forms deviate from slender in other respects, as will be discussed.

The 45 : 15 : 4 segregation is demonstrated by the F_2 in the subjoined figure. Here the three phenotypes are plainly distinguishable

TABLE 51.

Recombination

and the theoretical

Le Le La La Lb Lb
tall

Gametes ♀ / ♂	Le La Lb	Le La lb	Le la Lb	le La Lb
Le La Lb	Le Le La La Lb Lb. tall 1 : 0 : 0 1	Le Le La La Lb lb. tall 1 : 0 : 0 2	Le Le La la Lb Lb. tall 1 : 0 : 0 3	Le le La La tall 3 : 1 :
Le La lb	Le Le La La Lb lb. tall 1 : 0 : 0 9	Le Le La La lb lb. tall 1 : 0 : 0 10	Le Le La la Lb lb. tall 15 : 0 : 1 11	Le le La La tall 3 : 1 :
Le la Lb	Le Le La la Lb Lb. tall 1 : 0 : 0 17	Le Le La la Lb lb. tall 15 : 0 : 1 18	Le Le la la Lb Lb. tall 1 : 0 : 0 19	Le le La la tall 3 : 1 :
le La Lb	Le le La La Lb Lb. tall 3 : 1 : 0 25	Le le La La Lb lb. tall 3 : 1 : 0 26	Le le La la Lb Lb. tall 3 : 1 : 0 27	le le La La short 0 : 1 :
Le la lb	Le Le La la Lb lb. tall 15 : 0 : 1 33	Le Le La la lb lb. tall 3 : 0 : 1 34	Le Le la la Lb lb. tall 3 : 0 : 1 35	Le le La la tall 45 : 15 :
le La lb	Le le La La Lb lb. tall 3 : 1 : 0 41	Le le La La lb lb. tall 3 : 1 : 0 42	Le le La la Lb lb. tall 45 : 15 : 4 43	le le La La short 0 : 1 :
le la Lb	Le le La la Lb Lb. tall 3 : 1 : 0 49	Le le La la Lb lb. tall 45 : 15 : 4 50	Le le la la Lb Lb. tall 3 : 1 : 0 51	le le La la short 0 : 1 :
le la lb	Le le La la Lb lb. tall 45 : 15 : 4 57	Le le La la lb lb. tall 9 : 3 : 4 58	Le le la la Lb lb. tall 9 : 3 : 4 59	le le La la short 0 : 15 :

all × slender

tall : short : slender

le le la la lb lb
slender

| la lb | le La lb | le la Lb | le la lb |

.a la Lb lb. tall : 0 : 1 5	Le le La La Lb lb. tall 3 : 1 : 0 6	Le le La la Lb Lb. tall 3 : 1 : 0 7	Le le La la Lb lb. tall 45 : 15 : 4 8
.a la lb lb. tall 0 : 1 13	Le le La La lb lb. tall 3 : 1 : 0 14	Le le La la lb lb. tall 45 : 15 : 4 15	Le le La la lb lb. tall 9 : 3 : 4 16
ı la Lb lb. :all 0 : 1 21	Le le La la Lb lb. tall 45 : 15 : 4 22	Le le la la Lb Lb. tall 3 : 1 : 0 23	Le le la la Lb lb. tall 9 : 3 : 4 24
ı la Lb lb. :all 15 : 4 29	le le La La Lb lb. short 0 : 1 : 0 30	le le La la Lb Lb. short 0 : 1 : 0 31	le le La la Lb lb. short 0 : 15 : 1 32
ı la lb lb. nder 0 : 1 37	Le le La la lb lb. tall 9 : 3 : 4 38	Le le la la Lb lb. tall 9 : 3 : 4 39	Le le la la lb lb. slender 0 : 0 : 1 40
ı la lb lb. :all 3 : 4 45	le le La La lb lb. short 0 : 1 : 0 46	le le La la lb lb. short 0 : 15 : 1 47	le le La la lb lb. short 0 : 3 : 1 48
la Lb lb. all 3 : 4 53	le le La la Lb lb. short 0 : 15 : 1 54	le le la la Lb Lb. short 0 : 1 : 0 55	le le la la Lb lb. short 0 : 3 : 1 56
la lb lb. nder 0 : 1 61	le le La la lb lb. short 0 : 3 : 1 62	le le la la Lb lb. short . 0 : 3 : 1 63	le le la la lb lb. slender 0 : 0 : 1 64

as 3 layers owing to their difference in height. 7 slender plants were observed, 74 tall plants and 24 short plants.

The above F_2 has been further developed in the recombination

Fig. 4. F_2 tall × slender.

square shown in table 51, while in this scheme the segregations into tall : short : slender expected in the F_3 have likewise been given. It appears from the scheme that all *Le* forms belong to the tall type,

TABLE 52

F 3					F 2		
tall :	short :	slender	table	theor. on 64	observed	theor. expected
1 :	0 :	0		7	26	21.875
0 :	1 :	0		7	27	21.875
0 :	0 :	1		4	15	12.500
3 :	1 :	0	53	14	40	43.750
3 :	0 :	1	54	4	12	12.500
0 :	3 :	1	55	4	14	12.500
15 :	0 :	1	56	4	9	12.500
0 :	15 :	1	57	4	10	12.500
9 :	3 :	4	58	8	26	25.000
45 :	15 :	4	59	8	21	25.000
total of the F_3 generations					200	200.000

TABLE 53.   1928   40 F_3's of the cross 205.1 × slender

No.	tall	short	slender	total	theor.	3 : 1	m	D/m
12	230	75		305	228.8	76.2	7.58	0.158
19	16	8		24	18.0	6.0	2.00	1.000
27	96	38		134	100.5	33.5	4.90	0.918
33	184	88		272	204.0	68.0	6.78	2.949
35	22	16		38	28.5	9.5	2.35	2.766
41	174	42		216	162.0	54.0	6.60	1.818
42	288	127		415	311.3	103.7	8.49	2.744
43	229	106		335	251.3	83.7	7.57	2.946
44	139	65		204	153.0	51.0	5.90	2.373
45	404	131		535	401.3	133.7	10.05	0.269
46	182	44		226	169.5	56.5	6.75	1.852
47	296	96		392	294.0	98.0	8.60	0.233
52	164	73		237	177.8	59.2	6.40	2.156
57	178	53		231	173.3	57.7	6.67	0.705
73	193	85		278	208.5	69.5	6.95	2.230
81	300	131		431	323.3	107.7	8.66	2.691
84	257	94		351	263.3	87.7	8.02	0.785
85	281	123		404	303.0	101.0	8.38	2.625
90	308	105		413	309.8	103.2	8.77	0.205
91	135	63		198	148.5	49.5	5.81	2.323
92	259	117		376	282.0	94.0	8.05	2.857
99	480	140		620	465.0	155.0	10.95	1.370
103	85	27		112	84.0	28.0	4.61	0.217
107	338	117		455	341.3	113.7	9.19	0.359
122	123	61		184	138.0	46.0	5.55	2.703
131	82	28		110	82.5	27.5	4.53	0.110
139	328	114		442	331.5	110.5	9.06	0.386
147	295	107		402	301.5	100.5	8.59	0.757
148	203	70		273	204.8	68.2	7.12	0.253
149	182	84		266	199.5	66.5	6.75	2.593
153	216	55		271	203.3	67.7	7.35	1.728
156	90	24		114	85.5	28.5	4.74	0.949
162	138	67		205	153.8	51.2	5.87	2.691
170	23	15		38	28.5	9.5	2.40	2.292
171	216	88		304	228.0	76.0	7.35	1.633
178a	55	23		78	58.5	19.5	3.71	0.943
180	67	24		91	68.3	22.7	4.09	0.318
185	473	186		659	494.3	164.7	10.87	1.959
188	245	103		348	261.0	87.0	7.83	2.043
197	367	121		488	366.0	122.0	9.58	0.104
total				11475				

TABLE 54.   1928   12 F_3's of the cross 205.1 × slender

No.	tall	short	slender	total	theor. 3	: 1	m	D/m
1	99		32	131	98.3	32.7	4.97	0.141
13	84		28	112	84.0	28.0	4.58	0.000
51	289		95	384	288.0	96.0	8.50	0.118
62	171		47	218	163.5	54.5	6.54	1.147
79	84		35	119	89.3	29.7	4.58	1.157
114	190		59	249	186.8	62.2	6.89	0.464
115	120		42	162	121.5	40.5	5.48	0.274
121	264		79	343	257.3	85.7	8.12	0.825
127	595		187	782	586.5	195.5	12.19	0.697
135	331		106	437	327.8	109.2	9.10	0.352
190	342		113	455	341.3	113.7	9.25	0.076
205	237		81	318	238.5	79.5	7.70	0.195
total				3710				

TABLE 55.   1928   14 F_3's of the cross 205.1 × slender

No.	tall	short	slender	total	theor. 3	: 1	m	D/m
2		252	91	343	257.3	85.7	7.94	0.667
16		155	41	196	147.0	49.0	6.22	1.286
31		188	47	235	176.3	58.7	6.85	1.708
163		175	51	226	169.5	56.5	6.61	0.832
78		158	55	213	159.8	53.2	6.28	0.287
119		164	64	228	171.0	57.0	6.40	1.094
137		298	97	395	296.3	98.7	8.63	0.197
146		240	79	319	239.3	79.7	7.75	0.090
158		179	58	237	177.8	59.2	6.69	0.179
174		194	84	278	208.5	69.5	6.96	2.083
176		156	56	212	159.0	53.0	6.24	0.481
181		108	34	142	106.5	35.5	5.20	0.288
182		125	53	178	133.5	44.5	5.59	1.521
226		165	44	209	156.8	52.2	6.42	1.277
total				3411				

TABLE 56.   1928   9 F_3's of the cross 205.1 × slender

No.	tall	short	slender	total	theor. 15 : 1		m	D/m
7	285		23	308	288.8	19.2	4.22	0.900
21	343		23	366	343.1	22.9	4.63	0.022
29	79		4	83	77.8	5.2	2.22	0.541
32	141		13	154	144.4	9.6	2.97	1.145
49	265		21	286	268.1	17.9	4.07	0.762
58	282		20	302	283.1	18.9	4.20	0.262
95	118		10	128	120.0	8.0	2.72	0.735
109	479		30	509	477.2	31.8	5.47	0.329
221	450		32	482	451.9	30.1	5.30	0.358
total				2618				

TABLE 57.   1928   10 F_3's of the cross 205.1 × slender

No.	tall	short	slender	total	theor. 15 : 1		m	D/m
3		242	17	259	242.8	16.2	3.89	0.206
102		314	30	344	322.5	21.5	4.43	1.919
126		444	27	471	441.6	29.4	5.27	0.455
128		141	15	156	146.3	9.7	2.97	1.784
155		312	25	337	315.9	21.1	4.42	0.882
166		196	12	208	195.0	13.0	3.50	0.286
178b		468	30	498	466.9	31.1	5.41	0.203
194		286	19	305	285.9	19.1	4.23	0.024
198		325	17	342	320.6	21.4	4.51	0.976
202		190	18	208	195.0	13.0	3.45	1.449
total				3128				

except 37, 40 and 61, all of which are recessive for *La* and *Lb*. All the forms which are homozygous for *le* belong to the short type, except 64, which is recessive for *La* and *Lb*. To the slender type belong 37, 40, 61 and 64, all of them recessive for *La* and *Lb*. The F[3] segregations are given in table 52, the table that refers to each segregation, is placed, being mentioned behind it.

The observations of these F_3 were made on the germ plants. The total number of observations on the 200 F_3 cultures amounted to 51082. 26 F_3 cultures with a total of 7510 observations existed solely of tall plants, 27 F_3 cultures with a total of 6501 observations existed

Table 58. 1928 26 F_3's of the cross 205.1 × slender

No.	tall	short	slender	total	theor. 9	3	4		m			D/m	
17	99	20	47	166	93.3	31.1	41.5	6.58	5.23	5.45	0.866	2.122	1.009
25	84	42	31	157	88.3	29.4	39.2	6.06	4.64	5.61	0.710	2.715	1.461
324	48	15	27	90	50.6	16.9	22.5	4.58	3.75	3.97	0.568	0.507	1.133
34	90	42	60	192	108.0	36.0	48.0	6.27	5.30	5.74	2.871	1.132	2.091
40	108	34	55	197	110.8	36.9	49.2	6.87	5.53	5.96	0.407	0.524	0.973
56	165	60	58	283	159.2	53.1	70.7	8.50	6.47	7.50	0.682	1.066	1.693
64	169	58	79	306	172.1	57.4	76.5	8.60	6.82	7.53	0.360	0.088	0.332
65	78	15	47	140	78.7	26.2	35.0	5.84	4.84	4.82	0.120	2.314	2.489
66	86	40	27	153	86.1	28.7	38.2	6.13	4.60	5.61	0.016	2.456	1.996
67	122	62	71	255	143.4	47.8	63.8	7.30	6.01	6.78	2.931	2.363	1.062
70	21	5	.4	30	16.9	5.6	7.5	3.03	2.16	2.55	1.353	0.277	1.373
83	153	48	51	252	141.7	47.2	63.0	8.18	6.18	7.09	1.381	0.129	1.692
87	111	52	64	227	127.7	42.6	56.8	6.97	5.73	6.38	2.396	1.640	1.129
96	162	55	74	291	163.7	54.6	72.8	8.42	6.65	7.36	0.202	0.060	0.163
98	118	34	39	191	107.4	35.8	47.8	7.18	5.42	6.16	1.476	0.332	1.428
108	54	16	12	82	46.1	15.4	20.5	4.86	3.52	4.18	1.625	0.170	2.033
116	233	103	125	461	259.3	86.4	115.2	10.10	8.19	9.16	2.604	2.027	1.069
134	117	20	45	182	102.4	34.1	45.5	7.15	5.51	5.85	2.042	2.559	0.085
154	202	77	67	346	194.6	64.8	86.5	9.40	7.10	8.35	0.787	1.718	2.335
157	290	115	130	535	301.0	100.3	133.8	11.26	8.87	10.06	0.977	1.657	0.378
194	235	64	80	379	213.2	71.1	94.7	10.14	7.68	8.65	2.149	0.924	1.700
195	145	64	60	269	151.3	50.4	67.2	7.96	6.20	7.23	0.791	2.193	0.996
124	43	13	21	77	43.3	14.4	19.2	4.34	3.46	3.74	0.069	0.404	0.481
94	48	16	14	78	43.9	14.6	19.5	4.58	3.41	4.00	0.895	0.411	1.375
224	206	77	86	369	207.5	69.2	92.2	9.49	7.40	8.41	0.158	1.054	0.737
335	21	6	8	35	19.7	6.6	8.7	3.03	2.33	2.59	0.429	0.257	0.270
total				5743									

TABLE 59. 1928 21 F_3's of the cross 205.1 × slender

No.	tall	short	slender	total	theor. 45	15	4		m		D/m		
4	250	66	24	340	239.1	79.7	21.2	8.61	8.01	4.44	1.266	1.710	0.630
5	242	67	16	325	228.5	76.2	20.3	8.48	7.78	4.39	1.592	1.183	0.979
22	241	91	24	356	250.3	83.4	22.3	8.46	7.88	4.55	1.099	0.964	0.374
28	140	53	20	213	149.8	49.9	13.3	6.45	6.12	3.47	1.519	0.507	1.931
39	260	100	39	399	280.5	93.5	24.9	8.79	8.37	4.74	2.332	0.777	2.975
50	131	44	9	184	129.4	43.1	11.5	6.24	5.73	3.31	0.256	0.157	0.755
53	162	69	10	241	169.4	56.5	15.1	6.93	6.35	3.80	1.068	1.968	1.342
55	77	15	6	98	68.9	23.0	6.1	4.78	4.41	2.40	1.694	1.814	0.042
59	115	50	16	181	127.3	42.4	11.3	5.84	5.54	3.21	2.106	1.372	1.464
68	286	117	24	427	300.2	100.1	26.7	9.21	8.52	5.02	1.542	1.984	0.538
71	227	100	27	354	248.9	83.0	22.1	8.21	7.71	4.52	2.667	2.205	1.084
75	310	105	35	450	316.4	105.5	28.1	9.59	8.99	5.09	0.667	0.056	1.355
77	213	45	18	276	194.0	64.7	17.2	7.95	7.36	4.02	2.389	2.677	0.199
97	270	111	27	408	286.9	95.6	25.5	8.95	8.34	4.88	1.888	1.846	0.307
132	173	60	23	256	180.0	60.0	16.0	7.17	6.78	3.82	0.976	0.000	1.832
152	318	102	35	455	319.9	106.6	28.4	9.72	9.09	5.12	0.195	0.506	1.289
173	93	25	5	123	86.5	28.8	7.7	5.25	4.79	2.72	1.238	0.793	0.993
208	235	75	24	334	234.8	78.3	20.9	8.35	7.79	4.40	0.024	0.424	0.705
219	116	40	9	165	116.0	38.7	10.3	5.87	5.41	3.12	0.000	0.240	0.416
225	142	35	10	187	131.5	43.8	11.7	6.49	5.97	3.32	1.618	1.474	0.512
319	74	22	2	98	68.9	23.0	6.1	4.69	4.22	2.45	1.087	0.237	1.673
total				5870									

solely of short plants, while 15 F_3 cultures with a total of 1116 observations existed solely of slender plants. For briefness' sake the individual observations are not mentioned. The F_3 cultures in which segregations appeared occur in the tables 53—59.

The F_3 observations mentioned in the above tables agree with the theoretical expectation, as was given in the recombination square. Considerable departures are not observed. The objections SIRKS (70, 1929, p. 806) raised against various cases of polymeric factors, therefore, do not hold good for the factors *La* and *Lb*.

§ 4.   *A new case of polymeric growth inhibiting factors*

The investigation into the distribution of slender factors in the autumn of 1930 put me on the track of a new case of polymeric (multiple) factors. Among the F_2's of the various crosses with short 201.1 there occurred 2 which did not segregate into 15 short: 1 slender, but gave a 15 : 1 ratio for short and a somewhat taller type, which I did not notice before; this has been called short-2. The data appear in the subjoined tables.

TABLE 60. F_2 generation short 201.1 × short extra gekruiste
MANSHOLT

year	No.	short	short-2	total
1930	35.1	79	5	84
,,	35.2	392	29	421
total		471	34	505

theor. 15 : 1   473.5   31.5
m 5.42   D/m 0.461

TABLE 61. F_2 generation short 201.1 × short Unica

year	No.	short	short 2	total
1930	33.1	379	24	403
,,	33.2	173	11	184
total		552	35	587

theor. 15 : 1   550.3   36.7
m 5.87   D/m 0.290

From the 15 : 1 ratio it may be inferred that the short-2 form is double recessive for 2 factors, of which either of the parents possesses one. Just as in the case of the crypto dwarfs and the slender plants we may speak here of growth inhibiting factors. For the two new factors I chose the symbols *Lc* and *Ld*. As to 201.1 I assume that it possesses the factors *Lc Lc ld ld*, while in Unica and in the short extra gekruiste MANSHOLT the factors *lc lc Ld Ld* occur. The short-2 form has the constitution *lc lc ld ld*.

From the above we may conclude to a relationship between the Unica and the extra gekruiste MANSHOLT pea. On further inquiry it appeared that the Unica pea descends from a cross between the extra gekruiste MANSHOLT pea and the Wonder of Amsterdam. The relationship concluded from the investigation, therefore, was confirmed by this information.

Further investigation will have to teach in how far one of the slender factors (*La* or *Lb*) is identical to one of the factors *Lc* or *Ld*.

The germplants of the short-2 type somewhat resemble the wild pea (*Pisum elatius*); they differ from it, however, in the fact that the more vigorous growth does not continue as in *Pisum elatius*. Of the short-2 plants only the germ plants were observed in the field; in the autumn of 1930 they were transferred to the hothouse.

The short-2 form will be further examined and the genetical constitution for the other factors will be determined. Fig. 3 illustrates the various forms. From left to right we see slender — tall — crypto dwarf — short-2 — short.

From this it appears that a distinction between tall and crypto dwarf is impossible in this germ plant stage, so that from the F_2's in which the two types occur, mature plants must be grown.

Both for crypto dwarf and short-2 it may be inferred that they are recessive in crosses with normal short. The appearance of taller forms after crossing of short forms is interesting with respect to the investigation into the length factors in the short forms from practice, because it gives us some insight into the phenomenon that has been observed in agriculture for years, viz. the decline of a crop of peas owing to an increasing number of taller forms ('springers'), according as the culture has been cultivated longer without selection.

While the short plants in the field attain a length of 50 to 60 cm, the 'springers' are 80 to 90 cm tall. The difference in length is chiefly due to the somewhat longer internodes. The average number of the short peas is 16, of the 'springers' 18.

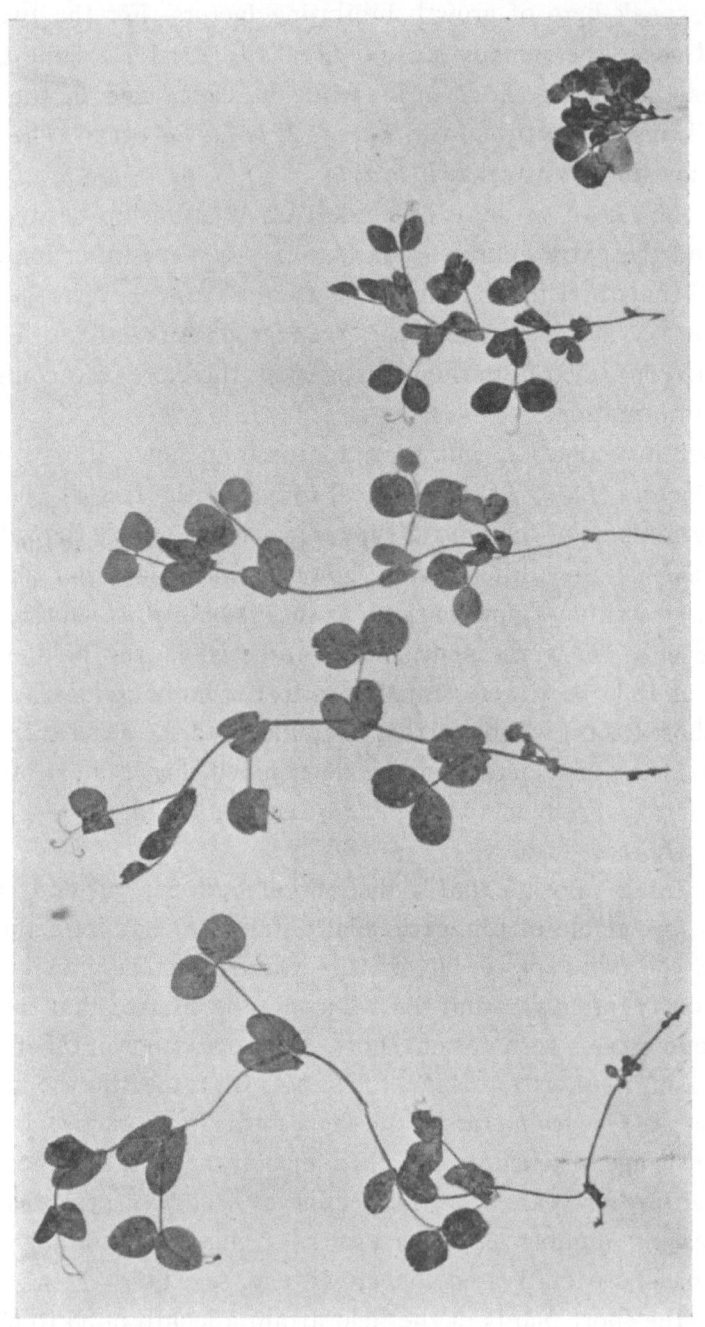

Fig. 5. Germ plants of slender, tall, crypto dwarf, short-2, short.

The short plants averagely bear 31 seeds, the 'springers' averagely 50 seeds. For the present these values may give an impression of the differences in type between the short forms and the 'springers'. About the genotypical difference, however, nothing is known. In the summer of 1930 I gathered material that is to serve for further investigation into this question. Very probably the decline off of a crop, due to a constant increase of taller forms, is based upon the heterozygotism of the original form. If this is correct, it will have to be proved that the 'springers' are recessive with respect to the original type.

§ 5. *The genetical relation between crypto dwarf and slender*

RASMUSSON (67, 1927, p. 40) communicated that the crypto dwarf plants first occurred in the F_2 of a cross of 2 short plants. From the correspondence in germ plant type of the crypto dwarf and the slender I inferred that the two forms might possibly be identical (39, 1927, p. 497). This, however, is not the case. The rapid growth of the crypto dwarf germ plants does not continue, so that the full grown plants differ little from the short parents. The question now arises, whether there exists any genetical relation between crypto dwarf and slender.

The research on this subject is still in its initial stage. Observations have been made on the F_2 germ plants of the crosses 201.1 (*la la Lb Lb*) × crypto dwarf and crypto dwarf × slender (*la la lb lb*). The data on the first mentioned cross are given in the subjoined table. From this it appears that no slender plants occurred, so that from this F_2 it may be inferred that the recessive factor *lb* does not occur in the

TABLE 62. F_2 generation of the cross 201.1 × crypto dwarf

year	No.	short	crypto dwarf	total
1929	88.1	54	16	70
,,	89.3	110	34	144
total		164	50	214

theor. 3 : 1   160.5   53.5
m 6.40   D/m 0.547

crypto dwarf. In order to trace whether the crypto dwarfs possess **the** factor *la*, they were crossed with slender.

TABLE 63.   F$_2$ generation of the cross crypto dwarf × slender

year	No.	crypto dwarf	slender	total
1929	82.2	66	20	86
1930	44.1	317	103	420
„	44.2	263	85	348
total		646	208	854

theor. 3 : 1   640.5   213.5

m 12.71   D/m 0.433

From this cross it may be concluded that the crypto dwarf and the slender differ in one factor. As this is not the *lb*, as has been demonstrated above, *la* may be concluded to. Thus the *la* factor has been demonstrated for the second time.

Moreover it may be inferred from the segregation into 3 short : 1 crypto dwarf in the F$_2$ short 201.1 × crypto dwarf (*cry$_1$ cry$_1$ cry$_2$ cry$_2$*) that short 201.1 is recessive for 1 of the *Cry* factors.

From the above it may be inferred that *la* will be identical with one of the *cry* factors, whether this is *cry$_1$* or *cry$_2$* has to be further investigated. On the priority of the *Cry*-signatures *La* and *la* will be dropped. Accordingly the formula of short 201.1 will be *le le Lb Lb cry$_1$ cry$_1$ Cry$_2$ Cry$_2$* or *le le Lb Lb Cry$_1$ Cry$_1$ cry$_2$ cry$_2$*, while crypto dwarf possesses the formula *le le Lb Lb cry$_1$ cry$_1$ cry$_2$ cry$_2$*. As long as it is not known with which *Cry* factor *La* is identical, I shall continue using the symbol *La*.

The F$_1$ crypto dwarf × 201.1 was short, the F$_1$ crypto dwarf × slender, on the other hand, stood between the two parents and corresponded in type with tall in spite of the fact that both crypto dwarf and slender are recessive for *Le*. Here again an interesting question presents itself concerning the cooperation of the length factors.

I wish to express my sincere thanks to Dr. RASMUSSON for sending me material.

§ 6. *The distribution of the factors la and lb over various forms.*

A beginning has been made with a research on the distribution of these factors *la* and *lb*. For this purpose crosses were made in the first place between short 201.1 (*la la Lb Lb*) and other pure lines I had at my disposal, while in those cases in which with short 201.1 no slender plants were obtained in the F_2, crosses were made with slender. Because the short 202.1 (*La La lb lb*) is usually off flowering in a short time and is moreover a form that has to be emasculated in a very young stage, it was little used.

In those cases in which by crossing with short 201.1 (*la la Lb Lb*) a 15 : 1 segregation was obtained it might be inferred that the studied form possessed the factors *La La lb lb*, if, on the other hand, no slender plants appeared, it proved that the factors *Lb Lb* occurred. In this latter case it had to be settled through crossing with slender (*la la lb lb*) whether the factor *La* or the factor *la* belonged to the genotype. In the former case a 15 : 1 segregation would occur, in the latter a 3 : 1 segregation.

First of all I mention the cases in which after crossing short 201.1 (*la la Lb Lb*) a 15 : 1 segregation was obtained. This is the case after the crossing of short 201.1 × schokker-1. The data have been mentioned in the subjoined table. From this it may be deduced that

TABLE 64.   Segregation into short and slender of the cross
schokker-1 × 201.1

year	No.	short	slender	total
1928	253	53	30	560
,,	255	446	32	478
total		976	62	1038
theor. 15 : 1		973.1	64.9	
		m 7.81	D/m 0.371	

schokker-1 possesses the factors *lb lb*. Besides schokker-1 I also had schokker-2 at my disposal, which cannot be distinguished from schokker-1 and just like schokker-1 was procured by Dr. MANSHOLT, so that I presume that the two forms are identical. The F_2 of schokker-2 and short 201.1 appears in the subjoined table. From this it may be inferred, that schokker-2 possesses the same factors as schokker-1.

Besides those above-mentioned I had 2 more schokkers left, viz. the 204.1 form already mentioned in the previous publication (39, 1927, p. 496) and the Glory schokker, supplied by Ir. KOOPMAN of Zierikzee. Observations on the cross of 204.1 and the short 201.1

TABLE 65.   Segregation into short and slender of the cross
201.1 × schokker-2

year	No.	short	slender	total
1930	32.1	294	18	312
,,	32.2	240	17	257
total		534	35	569

theor. 15 : 1   533.4   35.6
    m 5.78   D/m 0.104

form are mentioned in table 84 in an other connection. They confirm
the 15: 1 segregation previously observed; 204.1 therefore, has the factors *La La lb lb*. The observations on the cross 201.1 with the
Glory schokker appear in the subjoined table.

TABLE 66.   Segregation into short and slender of the cross
Glory schokker × 201.1

year	No.	short	slender	total
1930	36.1	328	25	353
,,	36.2	306	20	326
total		634	45	679

theor. 15 : 1   636.6   42.4
    m 6.29   D/m 0.413

From the 15 : 1 ratio and the occurrence of slender it appears that
in the Glory schokker the factor *Lb* occurs recessively.

Corresponding results I obtained by crossing the Zeeuwsche kroonerwt with short 201.1, as appears from the subjoined data.

TABLE 67.   Segregation into short and slender of the cross
201.1 × Zeeuwsche kroonerwt

year	No.	short	slender	total
1930	34.1	237	18	255
,,	34.2	153	9	162
total		390	27	417

theor. 15 : 1   390.9   26.1
    m 4.94   D/m 0.182

From the 15 : 1 segregation obtained it follows that the Zeeuwsche kroonerwt possesses the factors *La* and *lb*.

The Fletumer rozijnerwt and the purple patched pea derived from it, were likewise crossed with short 201.1. The result of the crossing short 201.1 with the Fletumer rozijnerwt is mentioned in the subjoined table. Here too if follows from the 15 : 1 ratio and the appearance of slender, that the Fletumer rozijnerwt possesses the

TABLE 68. Segregation into short and slender of the cross
Fletumer rozijnerwt × 201.1

year	No.	short	slender	total
1928	84.1	205	10	215
,,	84.2	267	18	285
,,	84.3	312	21	333
,,	84.4	220	14	234
total		1004	63	1067
theor. 15 : 1		1000.3	66.7	
		m 7.92	D/m 0.467	

factors *La La lb lb*. This was also the case with the purple patched, as appears from the subjoined table.

TABLE 69. Segregation into short and slender of the cross
201.1 × purple patched pea

year	No.	short	slender	total
1929	53.1	37	0	37
,,	53.2	89	3	92
,,	54.1	106	6	112
,,	54.2	105	7	112
,,	55.1	98	9	107
,,	55.2	90	6	96
,,	56.1	95	9	104
,,	57.1	53	4	57
total		673	44	717
theor. 15 : 1		672.2	44.8	
		m 6.49	D/m 0.123	

The agreement in hereditary constitution might be expected, because the purple patched occurred as a factor mutant for flower colour in the culture of Fletumer rozijnerwten and will, therefore, possess the same factors for length.

Besides in the forms mentioned I have been able to demonstrate the recessive factor *lb* in an other form, viz. in *Pisum Jomardi*, as appears from the subjoined data on the cross with short 201.1.

TABLE 70.   Segregation into tall, short and slender of the cross
201.1 × *Pisum Jomardi*

year	No.	tall	short	slender	total
1928	91.1	229	80	25	334
,,	91,2	447	150	37	634
,,	91.3	234	79	17	330
total		910	3.9	79	1298

theor. 45 : 15 : 4	912.6	304.2	81.1
m	16.43	15.22	8.73
D/m	0.158	0.315	0.241

From the above it appears that the recessive factor *lb* could be demonstrated for 202.1 (table 42) for slender (table 42) and further for schokker-1, schokker-2, 204.1, Glory schokker, Zeeuwsche kroonerwt, Fletumer rozijnerwt, purple patched pea and for *Pisum Jomardi*. In how far these *lb lb* forms are related will be further discussed.

In the following table a survey is given of the forms which crossed with short 201.1 did not give any slender plants in the F$_2$ and which, therefore, must possess the factor *Lb*. The totals of these observations occur in the subjoined table 71.

To the above forms which after crossing with 201.1 give no slender in the F$_2$, the previously mentioned 203, a popular short pea (39, 1927, p. 494) also belongs.

Further there belongs to the above forms the previously described tall 205 (39, 1927, p. 494), about which supplementary data were given in table 50, confirming the data given before.

As has been mentioned, it was concluded from the non appearance of slender plants in the above crosses with short 201.1 (*la la Lb Lb*), that the forms examined possess the factor *Lb*. As the factors might

TABLE 71. F$_2$ generations of crosses with 201.1 in which no
slender plants segregated

year				total
1927	paarse capucijner	×	201.1	143
,,	blauwpeul	,,	201.1	95
,,	peul	,,	201.1	60
1928	201.1	,,	langstroo acacia	85
,,	*Pisum quadratum*	,,	201.1	1546
,,	geelkelk	,,	201.1	365
,,	*Pisum elatius*	,,	201.1	2724
,,	apple	,,	201.1	502
,,	201.1	,,	vroege Hollandsche capucijner	501
,,	201.1	,,	rogue	706
,,	Alaska	,,	201.1	1282
,,	201.1	,,	recessive yellow	601
,,	paarse capucijner	,,	201.1	558
,,	rose 32	,,	201.1	468
,,	fasciated 85	,,	201.1	654
,,	201.1	,,	stamcapucijner	300
1929	33 little bloom	,,	201.1	2340
,,	201.1	,,	crypto dwarf	214
,,	light purple	,,	201.1	236
,,	201.1	,,	new rose	146
,,	crème 38	,,	201.1	165
,,	rose 11	,,	201.1	112
1930	Solo	,,	201.1	502
,,	Unica	,,	201.1	670
,,	extra gekr. *Mansholt*	,,	201.1	503
,,	1 M	,,	201.1	205

be either *La La Lb Lb* or *la la Lb Lb*, it was traced through
crossing with slender (*la la lb lb*) whether the *La* factor is present or not.

First of all I mention the cases in which after crossing with slender
a 15 : 1 ratio occurred. The following F$_2$ generations (tables 72-79)
were observed.

Further observations were made on the F$_2$ gathered from the cross
tall × slender, in which likewise a 15 : 1 segregation occurred, which
was discussed in an other connection (table 50).

From the 15 : 1 ratio it may be deduced that the *La* factor occurs
in the above forms.

TABLE 72. F_2 generation of the cross short 203.1 × slender

year	No.	short	slender	total
1927	33.1-35.2	448	25	473
1928	246	282	24	306
,,	263	195	18	213
,,	285	324	30	354
,,	289	238	15	253
,,	305	213	11	224
1929	87.1	300	21	321
total		2000	144	2144
theor. 15 : 1			2010	134

m 11.18   D/m 0.894

TABLE 73. F_2 generation of the cross slender × vale capucijner

year	No.	no slender	slender	total
1929	84.1	46	1	47
,,	84.2	96	8	104
total		142	9	151
theor. 15 : 1		141.6	9.4	

m 2.98   D/m 0.134

TABLE 74. F_2 generation of the cross paarse capucijner × slender

year	No.	no slender	slender	total
1928	297	367	28	395
,,	303	155	8	163
,,	413	244	21	265
total		766	57	823
theor. 15 : 1		771.6	51.4	

m 6.92   D/m 0.809

TABLE 75. F_2 generation of the cross slender × rogue

year	No.	no slender	slender	total
1928	227	135	7	142
,,	235	90	7	97
,,	280	164	10	174
total		389	24	413

theor. 15 : 1 387.2 25.8

m 4.93 D/m 0.365

TABLE 76. F_2 generation of the cross slender × fasciated 31

year	No.	no slender	slender	total
1929	83.1	345	22	367
,,	83.2	201	13	214
1930	83.3	233	15	248
total		779	50	829

theor. 15 : 1 777.2 51.8

m 6.98 D/m 0.258

TABLE 77. F_2 generation of the cross slender × 'springer' Unica

year	No.	short	slender	total
1930	42.1	275	18	293
,,	42.2	321	22	343
total		596	40	636

theor. 15 : 1 596.3 39.7

m 6.10 D/m 0.049

TABLE 78. F_2 generation of the cross slender × Early Giant

year	No.	no slender	slender	total
1929	85	23	—	23

theor. 15 : 1 21.6 1.4

m 1.2 D/m 1.167

TABLE 79. F_2 generation of the cross slender $\times$ acacia

year	No.	no slender	slender	total
1930	43.1	456	29	485

theor. 15 : 1   454.7   30.3
      m 5.34   D/m 0.243

A 3 : 1 ratio I obtained after crossing the Nunhem peul with slender as appears from the subjoined table.

TABLE 80. F_2 generation of the cross slender $\times$ Nunhem peul

year	No.	no slender	slender	total
1929	86	26	7	33

theor. 3 : 1   24.75   8.25
      m 2.55   D/m 0.490

From this it follows that the Nunhem peul possesses one of the slender factors, either *la* or *lb*. Through crossing with 201.1 this will be further investigated.

A 3 : 1 segregation I also found in the cross slender $\times$ crypto dwarf (table 63). From this F_2 may also be inferred that crypto dwarf possesses one of the slender factors, while from the cross crypto dwarf $\times$ short 201.1 (table 62), it was inferred that crypto dwarf possesses *Lb*. From this it follows that *la* will be the slender factor occurring in crypto dwarf. This proves at the same time that besides short 201.1 there is a second form with the factor *la*.

On a cross short (*La La lb lb*) I possess the following data.

TABLE 81. F_2 generation of the cross short 202.1 $\times$ new rose

year	No.	no slender	slender
1929	52.1	236	—
„	52.2	189	—
	total	425	—

From the non appearance of slender plants in this F_2 it may be concluded that new rose possesses the factor *La*. It was also con-

TABLE 82. Summary of the forms examined

		table			table
205.1	*Le Le La La Lb Lb*	50	vale capucijner	*le le La La Lb Lb*	73
fasciated 31	*Le Le La La Lb Lb*	76	paarse capucijner	*le le La La Lb Lb*	74
rogue	*Le Le La La Lb Lb*	75	'springer' Unica	*le le La La Lb Lb*	77
Early Giant	*Le Le La La Lb Lb*	78	203.1	*le le La La Lb Lb*	72
new rose	*Le Le La La Lb Lb*	81	201.1	*le le la la Lb Lb*	42
acacia	*Le Le La La Lb Lb*	79	crypto dwarf	*le le la la Lb Lb*	63
Pisum Jomardi	*Le Le La La lb lb*	70	202.1	*le le La La lb lb*	42
Pisum quadratum	*Le Le Lb Lb*	71	204.1	*le le La La lb lb*	84
geelkelk	*Le Le Lb Lb*	„	schokker-1	*le le La La lb lb*	64
Pisum elatius	*Le Le Lb Lb*	„	schokker-2	*le le La La lb lb*	65
apple	*Le Le Lb Lb*	„	Zeeuwsche kroon-		
vroege Holland-			erwt	*le le La La lb lb*	67
sche capucijner	*Le Le Lb Lb*	„	Glory schokker	*le le La La lb lb*	66
Alaska	*Le Le Lb Lb*	„	Fletumer rozijn	*le le La La lb lb*	68
fasciated 85	*Le Le Lb Lb*	„	purple patched	*le le La La lb lb*	69
light purple	*Le Le Lb Lb*	„	extra kortstroo		
rose 11	*Le Le Lb Lb*	„	*Mansholt*	*le le La La Lb Lb*	71
Solo	*Le Le Lb Lb*	„	Unica	*le le La La Lb Lb*	„
1 M	*Le Le Lb Lb*	„	33 little bloom	*le le Lb Lb*	„
Nunhempeul	*Le Le la of lb*	80	stamcapucijner	*le le Lb Lb*	„
			créme	*le le Lb Lb*	„
			recessive yellow	*le le Lb Lb*	„
			slender	*le le la la lb lb*	42

cluded from the cross with short 201.1 that the factor *Lb* occurred.

To get a survey the forms examined have been arranged in the foregoing table according to the factors for length. To the left we find the *Le Le* forms, to the right the *le le* forms.

From the above table it appears that the *la* factor was demonstrated in 2 cases, while it may be inferred that one of the parents of crypto dwarf will also possess the *la* factor.

Further investigation will have to teach us, whether it is likely that 201.1 and crypto dwarf are related, or that the *la* factor originated in them irrespective of each other.

As regards the other slender factor relationship could be demonstrated in some cases. As appears from table 82 the *lb* factor could be demonstrated for slender, 202.1, 204.1, schokker-1, schokker-2, Fletumer rozijn, purple patched pea, Zeeuwsche kroonerwt, Glory

schokker, *Pisum Jomardi*. Most of these are related as has been indicated in the subjoined scheme.

TABLE 83.   Survey of the distribution of the hereditary factor *lb*

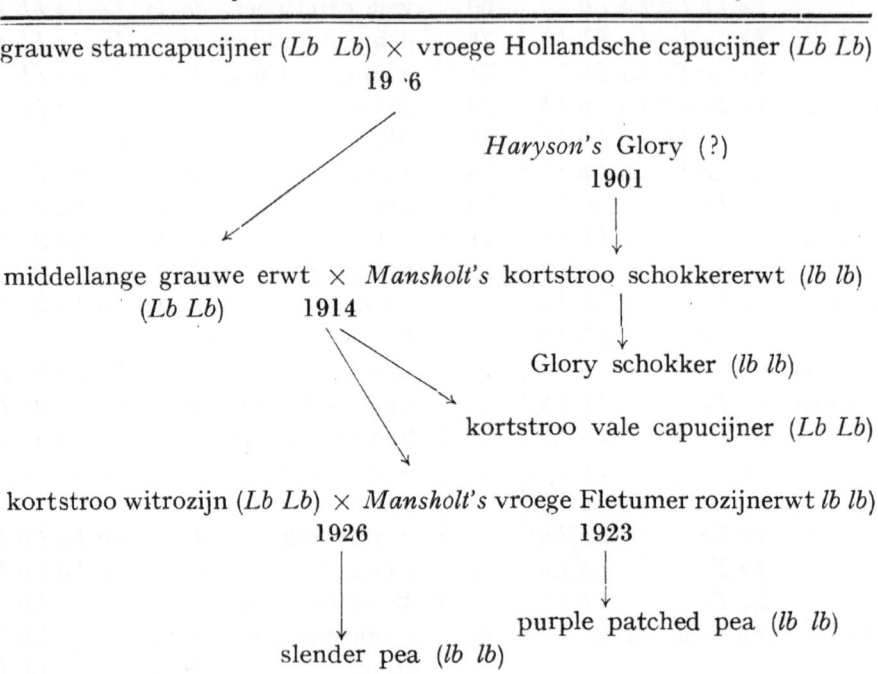

grauwe stamcapucijner (*Lb Lb*) × vroege Hollandsche capucijner (*Lb Lb*)
19 ·6

Haryson's Glory (?)
1901

middellange grauwe erwt × *Mansholt's* kortstroo schokkererwt (*lb lb*)
(*Lb Lb*)   1914

Glory schokker (*lb lb*)

kortstroo vale capucijner (*Lb Lb*)

kortstroo witrozijn (*Lb Lb*) × *Mansholt's* vroege Fletumer rozijnerwt *lb lb*)
1926   1923

purple patched pea (*lb lb*)

slender pea (*lb lb*)

From the survey it appears that the Fletumer rozijn was obtained
through crossing with the schokker. In a culture of Fletumer rozijn
the purple patched pea originated, the slender being obtained through
crossing short 201.1 and Fletumer rozijn. Schokker-1 and schokker-2
were grown by Dr. MANSHOLT and are presumably identical with the
schokker from which after crossing the Fletumer rozijnerwt is obtained. The Glory schokker is obtained from a culture schokker peas
of Dr. MANSHOLT, the descent of 204 is unknown to me; it has,
however, probably originated from a cross of a schokker. In how far
the Zeeuwsche kroonerwt is related with the schokkers I do not know.
It is a population that has been grown in Zeeland for many years.
Very likely, however, the Zeeuwsche kroonerwt will have arisen from
crossings with schokker peas. The *lb lb* factors of *Pisum Jomardi*
likewise make us conjecture a relationship between the schokkers and
this form. The investigation into this relationship is undoubtedly very
important with a view to the problem of the descent of the cultivated
populations.

From this research on the distribution of the slender factors appears the value these factors have with a view to the determination of the relationship between various forms in culture. For the very reason that the phenotype does not show whether a plant possesses the *lb* factor, the descent will be traceable on the method of genetical analysis.

Something similar I found for the new factors *Lc* and *Ld* (tables 60 and 61); since both the extra gekruiste MANSHOLT short and the Unica possess the *ld* factor, it is very likely that the two are closely related.

Such factors have the more value for the determination of relationship as they occur less frequently.

§ 7. *A case of absolute linkage*

In the genetic literature of the pea absolute linkage has been mentioned more than once. As will appear from the subjoined survey, however, absolute linkage has not been actually proved in any case.

First of all I mention the cases referring to absolute linkage of the factor *A* for flower colour with other factors.

TSCHERMAK (87, 1912, p. 232) explained the coincidence of a coloured leaf axil spot (*CC*), a coloured flower ($A_1 A_1$) and a coloured seedcoat (*Gc Gc*) as ascertained by MENDEL by assuming absolute linkage of the factors *C*, A_1 and *Gc*. WHITE (95, 1917, p. 171), however, pointed out that it is much simpler to regard these factors as one factor with many separate expressions. KAPPERT (49, 1923, p. 47) did not agree with this conception; his opinion was that we may assume absolute linkage of the factor *C* for the leaf axil colour with the factor *A* for flower colour. In a subsequent publication, however, (51, 1925, p. 583) he inferred from the genetic behaviour of the leaf axil colour in rose and in purple plants that it is the same factor that affects both the leaf axil and the flower.

WELLENSIEK (91, 1925, p. 380) considered the above factors as absolutely linked, whereas the TEDINS (85, 1926, p. 104) assumed that one and the same pleiotropic factor affects various parts of the plant. In a subsequent publication (the TEDINS and WELLENSIEK (84, 1925, p. 534) in which these investigators came to terms with respect to the symbolisation of the flower colour factors, WELLENSIEK likewise preferred to regard the factors considered in absolute linkage as in reality only one pleiotropic factor. Since that time the factor A has been maintained as the ground factor for colour in various parts of the plant.

At the same time it was accepted in various publications that A was in absolute linkage with the factor for indented seed L_1. The TEDINS and WELLENSIEK (84, 1925, p. 534) likewise considered this factor identical with A. L_1, however, is a factor for form determination and must not be considered identical with a factor for colour without further proofs. That, however, the assumption of the TEDINS and WELLENSIEK has been correct, may be inferred from the case of the purple patched peas, in which as was described, the seed form on the purple parts of the plant deviates from that on the spotted parts; the seed form of a purple branch is indented, whereas the seed-form of the spotted parts is smooth.

From the above it has appeared that the originally assumed absolute linkage of A with various factors for colour and with the factor for indented seed, is no longer maintained at present and instead of it a pleiotropic action has been assigned to A.

An other case in which there is question of absolute linkage is connected with the colour factor B. TEDIN (83, 1923, p. 41) surmises that TSCHERMAK's J (factor for dark brown seed colour) is identical or absolutely coupled with B. Very likely, however, the first is the case.

KAPPERT (49, 1923, p. 46) supposed absolute coupling of 1 of the factors for double leaf axil ring with the ground factor for colour, but in a subsequent publication (51, 1925, p. 586) he pointed out the possibility that we have to deal here with a case of multiple allelomorphism instead of absolute linkage. He did not come to a decision, however, concerning the two possibilities.

The TEDINS (85, 1926, p. 106) determined in the case of the double axil ring that here we have to deal with a case of multiple allelomorphism and chose the symbols $D^w - D - d$.

From the correlation LOCK found (57, 1905, p. 384) between seed form, leaf form and width of pod, WELLENSIEK (91, 1925, p. 435) inferred that this seems to hint at absolute coupling, since there appeared no new combinations of said characters.

In my cultures a case of absolute linkage occurred about which something was already published (39, 1927, p. 496). It could be shown that the factor lb was coupled with a factor w, causing a slight waxy

coating. In the F_2 of a cross between a short form containing this w
factor (*La La lb lb ww*) and a short form with a strong waxy coating
(*la la Lb Lb WW*) the expected 15 : 1 segregation was observed for
short: slender and it was at the same time established that all slender
plants belonged to the *ww* type. As only 39 slender plants were observed it was most important to repeat these crossings in order to
have a greater number of observations at our disposal. The totals of
40 F_2's occur in the subjoined table.

TABLE 84.   Segregation into short and slender of the cross
201.1 × 204.1

	short		slender		total
	glaucous	little bloom	glaucous	little bloom	
total of 40 F_2 's	5426	1417	—	439	7282
theor. 12 : 3 : 1	5461.2	1365.3	—	455.1	
m	36.8	33.2		20.7	
D/m	0.956	1.557		0.777	

A segregation took place according to the ratio 12 : 3 : 1.

For the 12 : 3 : 1 ratio from a cross of *la la Lb Lb WW* and *La La
lb lb ww* 2 explanations are possible. In the first place *lb* and *w* may
be identical and they must be replaced by one pleiotropic factor; in
the second place *lb* and *w* may be absolutely linked.

On the whole in similar cases the facts tell in favour of the assumption of pleiotropism if a relation can be established between the two
characters, either because both are founded on formation of pigment
or on inhibition of growth, etc. As the relation referred to can hardly
be established between the factor for slender and the factor for little
wax, there is no occasion to explain the above case by assuming the
pleiotropic action of one individual factor.

It follows that the explanation of absolute linkage ought to be
preferred and there are observations that corroborate this.

Besides *lb lb ww* and *Lb Lb ww* forms there are also *lb lb WW* forms,
as has been ascertained for the purple patched pea and as may be inferred from the occurrence of much wax in the Fletumer rozijn, the
schokker erwt, the Zeeuwsche kroonerwt and the Glory schokker. It

follows that besides *lb—w* and *Lb—W*, *lb—W* also occurs; whether *Lb—w* occurs I do not know.

It might be that the above combinations are based on multiple allelomorphism instead of on absolute linkage. This, however, is not probable, because totally different characters are concerned; the experiment of crossing cannot furnish proof.

From the above discussion it follows that in my opinion the 12 : 3 : 1 ratio should be explained by assuming absolute linkage of the factor *lb* and the factor *w*.

I do not know with which factor of WELLENSIEK's (92, 1928, p. 7) the *w* factor mentioned by me is identical. The investigation into the linkage of the factor for wax with one of the factors for slender will be supplemented by crossings with the forms of which WELLENSIEK determined the genetic constitution for wax.

§ 8. *The slender pea*

In a previous publication (39, 1927, p. 482) I described the most characteristic qualities of the slender peas. It appeared from these

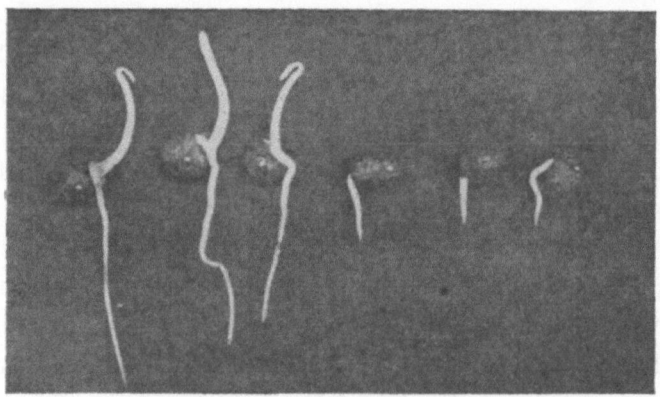

Fig. 6. Germ plants of the slender and the short pea.

that the slender pea is a fine instance of pleiotropism. Since that time some questions have been studied more fully.

The seeds of the slender peas germinate before those of the short plants, as is plainly demonstrated in the foregoing figure, where seeds from the short 201.1 and the slender pea sown at the same time have been represented. The seeds had been in a germinating pot in the hothouse for 3×24 hours, and were then removed from the pots. The young stem of the slender pea was $2\frac{1}{2}$—3 cm tall, the root $4\frac{1}{2}$—6 cm long; in short 201.1 on the other hand no growth of stem was to be noticed, while the roots were $1\frac{1}{2}$—2 cm long. It follows that also in the first stages of growth the slender pea develops considerably faster than the short 201.1. This also appears from figure 7,

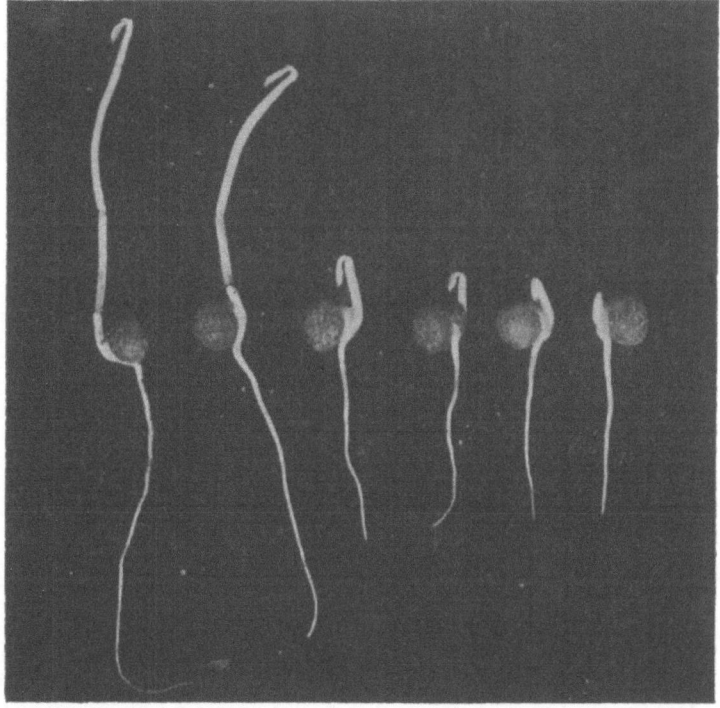

Fig. 7. Germ plants of the slender, the tall and the short pea.

which photograph was taken 4×24 hours after the sowing. Besides the slender and the short 201.1 two tall germ plants are shown in the centre of this figure.

From the investigation into the size of the epidermal cells it appeared that the epidermal cells of the slender plants compared with those of the short plants are very long. Of 4 greatly divergent types, viz. of the very short stamcapucijner, the short 202.1 form, the tall pea and the slender pea the epidermal cells have been represented in the following figures. For the research I chose the epidermis of the internode under the first flower, that is the centre of it. The average

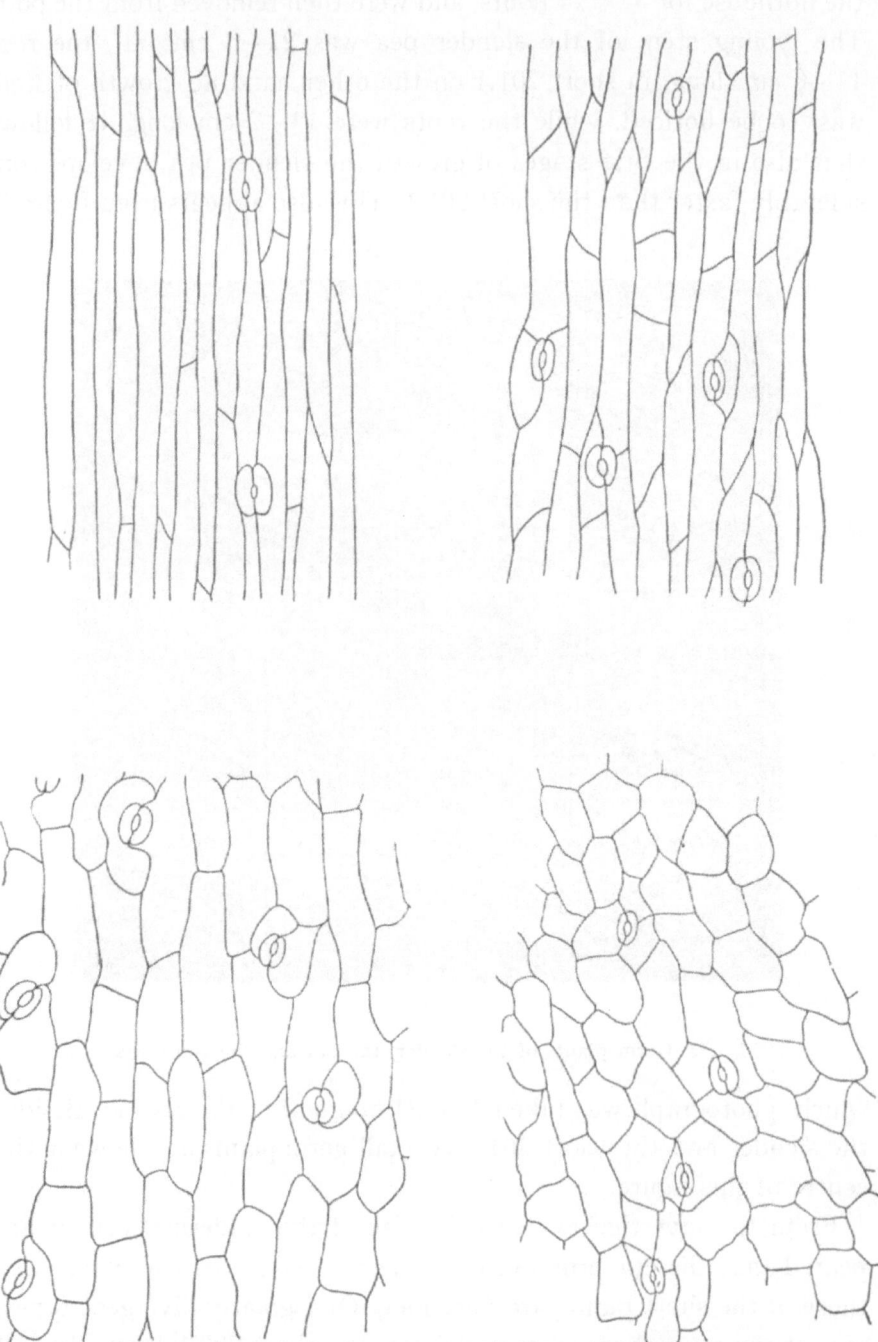

Fig. 8. Epidermal cells of the slender pea, the tall pea, the short 202.1 form and the
very short stamcapucijner. E = 140 ×.

Fig. 9. Photograph of the slender peas in the genetic experimental garden. For comparison a two meter high pair of steps was placed close behind the slender peas.

J. N. BAKKER phot.

cell lengths amounted respectively to 50, 125, 175 and 400 μ, the average cell widths to 40, 40, 30 and 20 μ.

Of the strongly deviating type several photographs were taken, one of which has been represented below. The tallest slender pea in this photo had a length of 4.35 m.

Part of the slender plants of the F_2 of the cross tall × slender (table 50) departed from the *le le* slender plants. In the first place it was inferred from these slender F_2 plants that all froms with the factor *Le* had no 2-leaved keel, but the keel had a more or less intermediary shape owing to the fact that the two keel leaflets only cohered at the bottom. The from the slender plants observed 55 keels consisted of 1 or 1—2 leaves and 17 of 2 leaves. This indicates a 3 : 1 ratio, in which the *Le* plants belong to the first, the *le le* plants to those with 2 leaves. *Le*, therefore, reduces the effect of the factors *la* and *lb*.

In a previous publication (39, 1927, p. 490) I already communicated something on the reduced fertility of the slender plants. Table 85 gives a survey of the number of seeds per pod, a total of 999 pods being examined.

TABLE 85.   Number of seeds per pod of the slender plants

number of seeds per pod	number of pods				% of the total
	1927	1928	1929	total	
0	77	120	113	310	31.0
1	21	108	43	172	17.2
2	33	97	42	172	17.2
3	30	98	39	167	16.7
4	32	41	33	106	10.6
5	10	19	21	50	5.0
6	7	1	12	20	2.0
7	0	1	1	2	0.2
total	210	485	304	999	

In normal cases about 5000 seeds may be expected from 999 pods. The total amount, however, was 1825, so that the fertility amounted to ± 36.5 %.

In order to obtain an insight into the cause of the reduced fertility it is material to trace how emasculated flowers behave. Previously (39, 1927, p. 490) I mentioned 10 cases in which the flowers were emasculated and allowed to develop without having been pollinated. No pods were formed. The emasculations were repeated in 1928 and after it appeared that from 19 emasculated flowers 8 pods were formed, the emasculations were once more repeated in 1929, various stages being chosen for this. 65 emasculated flowers formed 17 pods. Presumably I chose in the cases in which no pod was formed too young a stage for emasculation.

The question arose whether more seeds were formed after crossing than after self fertilization. In 1928 50 crossings were made from which 50 pods resulted, while in 1929 60 crossings were made from which 29 pods resulted. The subjoined table gives a survey.

TABLE 86. Number of seeds per pod from the crosses of the slender plants

number of seeds	number of pods	
	1928	1929
0	3	15
1	1	2
2	2	4
3	2	5
4	2	2
5	—	1
6	—	—
total	10	29

From a comparison of this table with the self pollinations in table 85 it may be inferred that crossed seeds do not possess more vitality than selfed ones.

From the emasculations it appears that pods are also formed without fertilization. Here therefore, a case presents itself of autonomic parthenocarpism (ENGLER, 31, 1926, p. 121). As the style is often hidden between the two keel leaflets, the possibility is created that no fertilization takes place. It is quite possible that part of the parthenocarpic pods develop without fertilization.

A large number of the parthenocarpic pods, however, are due to abortion of the seeds (stimulative parthenocarpism). The pods with a reduced number of seeds are also a result of abortion. The slender plants of the F_2 of the cross tall × slender, in which the factor Le occurred, showed much less abortion. Evidently the presence of Le reduces the effect of the factors la and lb.

Besides the case discussed in the slender peas there occur two cases of abnormal seed development in literature.

TEDIN (81, 1920, p. 73) ascertained that the violet plants produce aberrant seeds, in which the tracheal tissue of the hilum is entirely missing. Owing to this bad development a relatively large number of seeds were incapable of germinating.

In my cultures of these violet forms abortion in a very young stage was likewise frequent in all seeds in the pod, so that consequently a great many parthenocarpic pods originated. This was also the case in crosses with violet as mother, while in the reciprocal case no deviation occurred.

WELLENSIEK and KEYSER (93, 1928, p. 329) communicated a case of inherited abortion. The character demonstrates itself in a constriction of the pod in one or more places. In pods lacking the strong sclerenchymatous membrane at the innerside of the podwall, the abortion was not so easily observable as in pods where this membrane was well developed. This abortion is determined by a single dominant factor Q. Presumably there is no genetic relation with the seed abortion of the slender plants, inherited abortion being a dominant character and neither of the parents of the slender plant showing the said character.

Whether in the plants mentioned by WELLENSIEK and KEYSER pods occur of which all seeds are aborted is not mentioned.

From the above it follows that 3 cases of inherited abortion occur. The cases are genetically quite different. One case is connected with the pleiotropic effect of the slender factors, one case with the pleiotropic effect of the factors for violet flower colour, the 3rd case being controlled by a dominant factor, of which no further influence on any other part of the plant is known.

When this publication was in the press there appeared a paper by HåKANSSON (40, 1931, p. 39) on ringformation of the chromosomes in peas in which he mentions a case of semi-sterility of the amphibivalent plants.

CHAPTER IV

YELLOW VARIEGATION (STATUS ALBOMACULATUS)

In literature 2 cases of yellow variegation in the pea are mentioned. The data on its inheritance, however, are very incomplete.

KAJANUS (46, 1918, p. 83) communicated a case of constant yellow variegated plants. The occurrence of a few green plants in the progeny of the yellow variegated forms he attributed to spontaneous crossing with normal green plants. In a subsequent communication (KAJANUS, 47, 1927, p. 12) he stated that the F_1 of green × yellow variegated was green and that also its F_2 consisted of normal green plants. He inferred from this that presumably the yellow variegation is only inherited by the mother.

FRUWIRTH (34, 1920, p. 16) mentioned a case of inconstant yellow variegation. On the whole the plants were slightly yellow variegated. He ascertained that in the progeny of some green plants both green and yellow-variegated plants occurred and the yellow variegated individuals gave a progeny consisting of green or of green and yellow variegated plants. FRUWIRTH compared the hereditary behaviour with that of an ever-sporting variety.

In my cultures I observed once a case of white variegation and twice a case of yellow variegation. The first case concerned a normal green plant, on which a branch occurred which was nearly white-leaved at the foot and the leaflets of which were much reduced, as is shown in fig. 10 to the right. On the stipulae there likewise appeared white parts. From the white variegated part of this branch I got no seed; I did get some from the top which had passed into normal green. Its seeds just as the other seeds harvested from this plant gave a green progeny the next year.

The two cases in which yellow variegation occurred were independent of each other. Both times it was ascertained that on a normal green plant a yellow variegated branch occurred. One of these cases became the starting-point of a research on the hereditary behaviour of yellow variegation. The other case observed in the summer of 1930

Fig. 10.   Yellow variegated and white variegated leaves.

will be further investigated. This latter yellow variegated branch passed into yellow at the top.

In both cases plants were concerned from cultures I had been investigating for some years and in which variegation had never been observed.

The data mentioned below were gathered in 1928 from a yellow variegated branch occurring on a plant which was heterozygous for cotyledon colour and for flower colour ($I\ i\ A_2\ a$).

In 1929 I sowed out the seeds of the yellow variegated branch, of a yellow lateral branch of it and of the other green branches. These latter gave an exclusively green progeny, while the yellow branch yielded 3 pods with respectively 1, 2 and 4 seeds, giving normal green germ plants, which, however, yellowed and died off in a very young stage. The 35 seeds of the yellow variegated branch gave 35 green germ plants, 10 of which developed into normal green plants while 7 soon showed yellow variegated leaves and 18 yellowed and next died off. Both the yellowed germ plants and the slightly variegated germ plants of the same age have been represented in fig. 11. The upper

Fig. 11. Upper row slightly yellow variegated and lower row
yellowing germ plants.

row consists of 3 slightly yellow variegated germ plants, while in the lower row 4 germ plants of the same age occur which yellowed. Sometimes parts of the leaves of entire leaves are normal green; these germ plants should be taken as heavily yellow variegated plants. Among the strong yellow variegated plants some occurred that died off before the flowering stage. Just as the yellowing germ plants the yellow leaf spots were also originally green and then lost their chlorophyll. In the green cotyledons the phenomenon was not exhibited, but it did appear in the stipulae, the leaves and the sepals.

From the above it follows that in the progeny of the yellow variegated branch the normal green and the yellowing type occurred, while between the two extremes types occurred with more or less yellow variegation. On the yellow variegated plant the branches sometimes differed in the degree of yellow variegation, while moreover green and yellow branches occurred. In many cases a yellow variegated branch passed into green, in an exceptional case the yellow variegated branch passed into yellow.

While the yellow variegated branches gave a progeny of green, yellow variegated and yellowing and the ratio of them was shifted more to yellow variegated and yellow, according as the leaves showed a stronger degree of variegation, the green branches on the yellow variegated plant gave exclusively green, the yellow branches exclusively yellow, while for yellow variegated branches that passed into green or yellow the same held good.

From the crosses of strongly yellow variegated forms, of which I always chose flowers with a strongly yellow variegated calyx, with normal-green plants, it appeared that the 12 F_1 plants with yellow variegated as father were normal green and also in their progeny bred true for green, the crosses with strong yellow variegation as mother plant, on the other hand, gave the same picture as selfings of the variegated plant. Though a number of questions are still being studied, it may be determined that yellow variegation is not inherited with the pollen and can only be kept up by self-pollinating the variegated plant or by using it as mother-plant in crossings.

Of some of the yellow variegated plants from 1929 the seed was sown in 1930. The data are mentioned in the subjoined table 87.

In the autumn I also sowed out the seed of some plants from the crop 1930, on the germ plants of which the following observations were made (table 88).

In the variegation cultures a segregation appeared according to the cotyledon colour and the flower colour; the figures observed were conform with the expectation. From this it may be inferred that characters transmitted by the nucleus give normal segregations.

As already stated the F_1 plants of crosses of strongly yellow variegated × green resembled the selfing of strongly yellow variegated. Of F_1 No. 49 3 plants were cultivated, 2 of which yellowed as germ plants, while one being heavily yellow variegated died off before the flowering stage. Of F_1 No. 57 there were 4 plants moderately yellow variegated, while 2 yellowed. Of F_1 No. 50 2 were normal green and 2 yellow variegated. The 13 F_1 plants therefore, could be subdivided into 2 green, 7 yellow variegated and 4 yellowing ones; the 12 F_1 plants with strongly yellow variegated as father and normal green as mother,

TABLE 87. Self-pollination of variegated plants

No.	description 1929		1930			
			green	yellow varie-gated	yello-wing	total
7.1	branch 1	moderately yellow variegated	14	2	2	18
	branch 2	„ „	8	2	4	14
	branch 3	moderately yellow variegated pas-sing into green	16	1	1	18
	branch 4	green	27	—	—	27
2.2	branch 1	yellowing	—	—	7	7
	branch 2	strongly yellow va-riegated	19	9	25	53
3.2	branch 1	moderately yellow variegated	8	2	2	12
	branch 2	slightly yellow va-riegated, passing into green	24	1	—	25

TABLE 88. Self-pollination of variegated plants

No.	description 1930		1930			
			green	yellow varie-gated	yello-wing	total
5.4	branch 1	strongly yellow va-riegated	8	5	9	22
	branch 2	strongly yellow va-riegated	2	5	13	20
8.1	not branched	faintly yellow varie-gated	29	3	—	32
6.4	not branched	faintly yellow varie-gated	16	—	—	16

on the other hand, were all of them normal green. Though it is
desirable to increase the data by some crossings, especially with strong
yellow variegation as father, it may be inferred from the observations
that yellow variegation is not controlled by a hereditary factor in the
nucleus, but can only be transmitted from cell to cell by the maternal
plasm. It may likewise be ascertained that in the progeny by the side

of *albomaculata* forms, also *albina* occurs in green plants, which, there-fore, is a departure from the case of constant yellow variegation observed by KAJANUS.

Owing to the great resemblance in hereditary behaviour and in outward appearance with other cases of yellow variegation indicated in literature with *albomaculata* the above described yellow variegation was likewise called status albomaculatus.

CORRENS (26, 1928, p. 144) pointed out the difficulties to explain the inheritance of the *albomaculatus* type by assuming two kinds of plastids. Instead of it CORRENS assumed that the plasm can occur in different states. Externally this representation has points of resem-blance with the multiple allelomorphs; in how far there are internal points of resemblance is still unknown.

In agreement with CORRENS' representations I assume that in 1928 a plasmogenic mutation occurred (BAUR, 7, 1930, p. 311) in the normal green short pea. The plasm of the modified cell was labile. Since then this labile state of the embryonic cells has been maintained and has directly been transmitted from the cell to the cells originated from it by division. The labile plasm, however, passed in part of the cells into constant normal plasm, the cells formed from it by division also having normal plasm. In another part of the cells the plasm passed into constant 'ill', as the plasm of the yellowing cells is called by CORRENS (26, 1928, p. 144); and this plasm continued 'ill' also in the cells formed from it by division. So far as cells were concerned with chloroplasts the first mentioned were normal green while in the cells with diseased plasm the originally green chloroplasts yellowed. With regard to the ovules it should be assumed that they continue partly in the state of labile plasm, so that we may find here 3 types side by side: green — labile — 'ill'. The state of the plasm of the ovule determines the type that proceeds from it, irrespective of the plant used as father. An ovule with normal plasm gives normal green plants, with labile plasm it gives yellow variegated plants and with ill plasm yellowing plants. If the vegetation point consists exclusi-vely of cells with ill plasm, a yellowing branch will arise from it, while a green branch arises from a vegetation point consisting of cells with normal plasm.

The two cases mentioned in literature differ from each other and likewise from the case of yellow variegation described by me. KAJA-

NUS' data, however, are at variance, as already observed by WELLEN-
SIEK (91, 1925, p. 395). If his case of yellow variegation is indeed
constant and is only inherited maternally, it is theoretically impossible that through spontaneous cross-pollination with a green plant
there should occur green plants in the progeny. Most likely it will be
a case of pollution. If this is true, the 2nd case of constant maternal
inheritance, that is, the case of yellow variegation in *Pisum* described
by KAJANUS, corresponds with that of *Humulus*. (WINGE, 98, 1919,
p. 9). Further details of this highly important case, however, are
unknown to me.

Besides of yellow variegated peas cases are known of yellow plants
(RASMUSSON, 68, 1929, p. 611), of white plants (BATESON, 2, 1915,
p. 33; FRUWIRTH, 34, 1920, p. 9; RASMUSSON, 68, 1929, p. 611) and
of white variegated plants (SPRENGER, 74, 1916, p. 782). SPRENGER
assumes that white variegated inherits plasmatically. BATESON,
FRUWIRTH and RASMUSSON ascertained that white behaves as a
recessive character, inheriting monofactorially. To this factor WEL-
LENSIEK (91, 1925, p. 395) assigned the symbol *Wb*. The number of
white plants, however, was below the number theoretically expected.
The yellow and the white plants died as germ plants. RASMUSSON
called the factors for it lethal factors.

SCHERZ (69, 1927, p. 39) gave a summary of the cases of maternal
inheritance of variegation. Since then some new cases have been
described. In the subjoined table SCHERZ' list has been supplemented
and the cases of maternal inheritance of variegation known at present have been recorded.

TABLE 89.
Summary of the cases of maternal inheritance of variegation

II-*A-b*   (maternal inheritance in which by the side of *albomaculata*
there also occur *albina* and normal forms in the progeny)
Mirabilis (CORRENS, 1909)
Antirrhinum (BAUR, 1911; GAIRDNER and HALDANE, 1929)
Primula sinensis (BAUR, 1911; GREGORY, 1915)
Trifolium (KAJANUS, 1913)
Melandrium (SHULL, 1913)

  Pisum (white variegation, SPRENGER, 1916)
  Campanula (PELLEW, 1917)
  Urtica (CORRENS, 1919)
  Mesembryanthemum (CORRENS, 1919)
  Arabis (CORRENS, 1919)
  Aubretia (CORRENS, 1919)
  Oenothera (STOMPS, 1920)
  Mercurialis (CORRENS, 1920)
  Beta (STEHLÍK, 1921; MUNERATI, 1928)
  Hordeum (Sô, 1921)
  Stellaria (CORRENS, 1922)
  Senecio (CORRENS, 1922)
  Taraxacum (CORRENS, 1922)
  Hieracium (CORRENS, 1922)
  Chlorophytum (COLLINS, 1922)
  Zea Mays (ANDERSON, 1923)
  Mimulus (BROŽEK, 1923)
  Hydrangia (CHITTENDEN, 1926)
  Nicotiana (HONING, 1927)
  Oryza (KONDÔ, TAKEDA, FUJIMOTO, 1927; TAKEZAKI, 1922)
  Primula veris (CHATTAWAY and SNOW, 1929)
  Viola (CLAUSEN, 1929, intermediate type for II-*A-a*)
  Hosta (YASUI, 1929)
  Pisum (yellow variegation, DE HAAN, 1931)
II-*B-b* (maternal inheritance, constant *albomaculata*)
  Humulus (WINGE, 1917, 1919)
  Pisum (yellow variegation, KAJANUS, 1924)?

SUMMARY

Two hitherto unknown flower colours of *Pisum*, called apple and pinkish white, have been described and their heredity studied. Both differ from purple in 1 factor; these factors which are dominant in purple have been called A_p and A_m. Apple and pinkish white have been considered factor mutants. Crossing with apple resulted in 2 new flower colours: apple violet and apple rose, both differing from purple in 2 factors.

The relation between flower colour and leaf axil colour has been more fully discussed. The factor A_p does not affect the leaf axil, only the flower, whereas the factor A_m affects flower and leaf axil in a different way.

A connection has been established between the colour factors and the leaf colour in spring.

In a large culture of purple-flowering peas a purple patched form occurred. Purple patched consists of a mosaic of white and purple parts. Besides the other factors for purple the white parts have instead of A a factor A_2 arisen from it through mutation. The purple parts, however, do possess the factor A, A_2 being a labile factor, which can pass into A. Besides purple parts also purple dotted parts arise on the purple patched plant.

Purple patched is not constant; in the progeny there occur besides purple patched also purple, purple dotted and 'white', that is extremely faint purple patched.

The crosses of purple patched × white are reciprocally different. When purple patched is used as mother there also occur besides purple patched, purple and purple dotted F_1 plants; when on the other hand, purple patched is used as father, all F_1 plants are either purple or purple dotted, but never purple patched. In order to explain the genetic behaviour a working-hypothesis has been set up, based on the assumption that patching is due to the mutation $A_2 \rightarrow A$ in cells with labile plasm. If the labile plasm passes into normal plasm no mutation takes place and purple dotted arises. There is a fundamental difference between purple patched and purple dotted; the former is a mosaic of genetically different parts, the latter is a pattern of which the coloured and the colourless cells have the same hereditary constitution.

In the cultures a new form, crypto purple occurred, for which the factor A_1 has been established. The factors for purple, crypto purple, purple dotted and white form part of a series of multiple allelomorphs $A - A_1 - A_2 - a$.

The interaction of the two growth-inhibiting factors La and Lb and the growth-favouring factor Le has been investigated and the F_2 data have been confirmed by an extensive F_3.

A new case of two polymeric factors, called Lc and Ld has been determined; these factors also have a growth-inhibiting effect.

From the research into the genetic relation of the crypto dwarfs and the slender peas it has been concluded, that *La* is identical with 1 of the crypto dwarf factors.

An investigation has been made into the distribution of the factors *la* and *lb*. In the forms examined the *la* factor is very rare, while *lb* is demonstrated in various forms. Afterwards it appeared that forms possessing *lb* are related. From this follows the value of the factors *la* and *lb* for determining relationship in cases in which nothing is known about the descent. For the very reason that the hereditary constitution cannot be determined from the phenotype, the descent will be traceable with the aid of genetic analysis. This same holds good for the factors *lc* and *ld*.

The *lb* factor is absolutely linked with 1 of the factors for wax.

With respect to the slender pea it appears that without fertilization parthenocarpism can occur.

On one short plant a yellow variegated branch appeared. From its progeny it was inferred that yellow variegation corresponds with the status *albomaculatus* as known in other plants. The reciprocal crosses with normal green were different. Yellow variegation is not transmitted by the father. To explain the hereditary behaviour it has been adopted that the plasm of the variegated plant is in a labile state and that the green patches are due to the transition from the state of labile to that of normal plasm.

LITERATURE

1. ANDERSON, E. G., Maternal inheritance of chlorophyll in maize. Bot. Gaz. 76, 1923, p. 411.

2. BATESON, W. and PELLEW, C., On the genetics of 'rogues' among culinary peas (*Pisum sativum*). Journ. of Gen. 5, 1915—1916, p. 13.

3. BATESON, W., Segregation:. The Joseph Leidy Memorial Lecture of the University of Pennsylvania, 1922. Journ. of Gen. 16, 1926, p. 201.

4. BAUR, E., Untersuchungen über die Vererbung von Chromatophorenmerkmalen bei *Melandrium, Antirrhinum* und *Aquilegia*. Zeitschr. f. ind. Abst. und Vererb. 4, 1910—1911, p. 81.

5. BAUR, E., Einführung in die Vererbungslehre, Berlin, 1914.

6. BAUR, E., Untersuchungen über das Wesen, die Entstehung und die Vererbung von Rassenunterschieden bei *Antirrhinum majus*. Bibliotheca Genetica 4, 1924, p. 1.

7. BAUR, E., Einführung in die Vererbungslehre, Berlin, 1930.

8. BRAUN, A., Betrachtungen über die Erscheinung der Verjüngung in der Natur. Leipzig, 1851.

9. BROŽEK, A., Selektions- und Kreuzungsexperimente mit albomaculaten (weiszbunten) *Mimulus*-Rassen. Stud. Plant Physiol. Lab. Charles Univ. Praque 1, 1923, p. 43.

10. CAYLEY, D. M., 'Breaking' in tulips. Ann. appl. Biol. 15, 1928, p. 529.

11. CHATTAWAY, M. M. and SNOW, R., The genetics of a variegated primrose. Journ. of Gen. 21, 1929, p. 81.

12. CHITTENDEN, R. J., Studies in variegation. II. *Hydrangia* and *Pelargonium*: with notes on certain chimerical arrangements which involve sterility. Journ. of Gen. 16, 1926, p. 43.

13. CHITTENDEN, R. J., Vegetative segregation. Bibliogr. Genetica 3, 1927, p. 355.

14. CLAUSEN, J., Genetical and cytological investigations on *Viola tricolor* L. and *Viola arvensis* M u r r.. Hereditas 8, 1926—1927, p. 1.

15. CLAUSEN, J., Inheritance of variegation and of black flower colour in *Viola tricolor* L. Hereditas 13, 1929—1930, p. 342.

16. CLAUSEN, R. E., Inheritance in *Nicotiana tabacum*. 10. Carmine-coral variegation. Cytologia 1, 1929—1930, p. 358.

17. COLLINS, E. J., Variegation and its inheritance in *Chlorophytum elatum* and *C. comosum*. Journ. of Gen. 12, 1922, p. 1.

18. CORRENS, C., Ueber Bastardirungsversuche mit *Mirabilis*-Sippen. Erste Mittheilung. Ber. d. D. Bot. Ges. 20, 1902, p. 594.

19. CORRENS, C., Zur Kenntnis der scheinbar neuen Merkmale der Bastarde. Zweite Mitteilung über Bastardierungsversuche mit *Mirabilis*-Sippen. Ber. d. D. Bot. Ges. 23, 1905, p. 70.

20. CORRENS, C., Vererbungsversuche mit blass(gelb)grünen und buntblättrigen-Sippen bei *Mirabilis Jalapa, Urtica pilulifera* und *Lunaria annua*. Zeitschr. f. ind. Abst. und Vererb. 1, 1909, p. 291.

21. CORRENS, C., Der Übergang aus dem homozygotischen in einen heterozygotischen Zustand im selben Individuum bei buntblättrigen und gestreiftblühenden *Mirabilis* Sippen. Ber. d. D. Bot. Ges. 28, 1910, p. 418.

22. CORRENS, C., Vererbungsversuche mit buntblättrigen Sippen. 1. *Capsella Bursa pastoris albovariabilis* und *chlorina*. Sitz. Ber. d. Preuss. Akad. 1919, p. 585.

23. CORRENS, C., Vererbungsversuche mit buntblättrigen Sippen. II. Vier neue Typen bunter Periklinalchimären. Sitz. Ber. d. Preuss. Akad. 1919, p. 820.

24. CORRENS, C., Vererbungsversuche mit buntblättrigen Sippen. V. *Mercurialis annua versicolor* und *xantha*. Sitz. d. Preuss. Akad. 6, 1920, p. 232.

25. CORRENS, C., Vererbungsversuche mit buntblättrigen Sippen. VI. Einige neue Fälle von albomaculatio. Sitz. d. Preuss. Akad. 33, 1922, p. 460.

26. CORRENS, C., Über nichtmendelnde Vererbung. Verh. 5. intern. Kongr. f. Vererb. W. Leipzig, 1928, p. 131.

27. DARWIN, CH., The variation of plants and animals under domestication. London, 1868.

28. DARWIN, CH., The effects of cross- and self-fertilisation in the vegetable kingdom. London, 1876.

29. DAVENPORT, CH. B., Inheritance of stature. Genetics 2, 1917, p. 313.

30. DEMEREC, M., *Delphinium Ajacis*, Year book Carnegie Institution of Washington 25, 1926, p. 35.

31. ENGLER, A., Die natürlichen Pflanzenfamilien. 14a, Leipzig, 1926.

32. EYSTER, W. H., The mechanism of variegations. Verh. 5. intern. Kongr. für Vererbungsw., Leipzig, 1928, p. 666.

33. FEDOTOV, V. S., On the hereditary factors of flowercolour and of some other characters in the pea. Proc. of the U S S R Congress of Genetics, Plant- and Animal Breedings, vol. 2, 1930, p. 523.

34. FRUWIRTH, C., Neunzehn Jahre Geschichte einer reinen Linie der Futtererbse. Fühlings Landw. Zeitung 69, 1920, p. 1.

35. GAIRDNER, A. E. and HALDANE, J. B. S., A case of balanced lethal factors in *Antirrhinum majus*. Journ. of Gen. 21, 1929, p. 315.

36. GOVOROV, L. J., The peas of Afghanistan. Bull. of Appl. Bot. 19—2, 1928, p. 497.

37. GREGORY, R. P., Experiments with *Primula sinensis*. Journ. of Gen. 1, 1910—1911, p. 73.

38. GREGORY, R. P., On variegation in *Primula sinensis*. Journ. of Gen. 4, 1914—1915, p. 305.

39. HAAN, H. DE, Length factors in *Pisum*. Genetica 9, 1927, p. 481.

40. HÅKANSSON, A., Über Chromosomenverkettung in *Pisum*. Hereditas 15, 1931, p. 17.

41. HONING, J. A., Erblichkeitsuntersuchungen an Tabak. Genetica 9, 1927, p. 1.

42. IKENO, S., Studien über die Vererbung der Blütenfarbe bei *Portulaca grandiflora*. III. Mitt. Mosaicfarbe. Jap. Journ. of Botany 4, 1929, p. 189.

43. IMAI, Y., Genetic studies in Morning Glories. 15. On the eversporting behavior of cream flowers in *Pharbitis Nil*. Bot. Magaz. Tôkyô 39, 1925, p. 43.

44. IMAY, Y., The vegetative and seminal variations observed in the Japanese Morning Glory, with special reference to its evolution under cultivation. Journ. of the Coll. of Agric. Tôkyô 9, 1927, p. 223.

45. KAJANUS, B., Über einige vegetative Anomalien bei *Trifolium pratense* L. Zeitschr. f. ind. Abst. und Vererb. 9, 1913, p. 111.

46. KAJANUS, B., Über eine konstant gelbbunte *Pisum*-Rasse. Botaniska Notiser 1918, p. 83.

47. KAJANUS, B., Genetische Studien an *Pisum*. Zeitschr. f. Pfl. Zücht. 9, 1924, p. 1.

48. KAPPERT, H., Untersuchungen über den Merkmalskomplex glatte-runzelige Samenoberfläche bei der Erbse. Zeitschr. f. ind. Abst. und Vererb. 24, 1921, p. 185.

49. KAPPERT, H., Über ein neues einfach mendelndes Merkmal bei der Erbse. Ber. d. D. Bot. Ges. 41, 1923, p. 43.

50. KAPPERT, H., Über die Zahl der unabhängigen Merkmalsgruppen bei der Erbse. Zeitschr. f. ind. Abst. und Vererb. 36, 1925, p. 1.

51. KAPPERT, H., Über absolut gekoppelte Faktoren oder multiple Allelomorphe bei *Pisum*. Ber. d. D. Bot. Ges. 43, 1925, p. 582.

52. KAPPERT, H., Die Erblichkeitsverhältnisse der züchterisch wichtigen Eigenschaften der Gartenerbse. Der Züchter 1, 1929, p. 79.

53. KAZNOWSKY, L., Recherches sur le pois. Mém. d. l'Inst. nation. pol. d'écon. rurale à Pulawy 7, 1926.

54. KEEBLE, F. and PELLEW, C., The mode of inheritance of stature and time of flowering in peas (*Pisum sativum*). Journ. of Gen. 1, 1910—1911, p. 47.

55. KOJIMA, H., The inheritance of mosaic flower colour in a race of *Celosia cristata* L. Bot. Mag. Tôkyô 44, 1930, p. 328.

56. KONDÔ, M., TAKETA, M. and FUJIMOTO, S., Untersuchungen über die weissgestreifte Reispflanze. Ber. Ohara Inst. landw. Fortschr. 3, 1927, p. 291.

57. LOCK, R. H., Studies in plant breeding in the tropics. II. Experiments with peas. Ann. of the Royal Bot. Gard. Peradeniya 2, 1905, p. 357.

58. MARRYAT, D. C. E., Hybridisation experiments with *Mirabilis Jalapa*. Reports to the evolution committee of the Royal Society. Report 5, 1909, p. 32.

59. MENDEL, G., Versuche über Pflanzen-Hybriden. Verhandlungen des naturf. Verein. Brünn, 4, 1866, p. 3.

60. MILLER, P., Dictionaire des Jardiniers et des Cultivateurs. Bruxelles, tome 6, 1786, p. 16.

61. MUNERATI, O., L'Hérédité de l'Albinisme en *Beta vulgaris* L. Verh. 5. intern. Kongr. f. Vererb. W. 2, Leipzig, 1928, p. 1137.

62. NAUDIN, CH., Nouvelles recherches sur l'hybridité dans les végétaux (Conclusions du mémoire couronné. Ann. Sc. Nat. 4e série botanique, tome 19, 1863, p. 180.

63. NILSSON-EHLE, H., Kreuzungsuntersuchungen an Hafer und

Weizen. 2. Lunds Univ. Arsskrift, N. F., Afd. 2, Bd. 7, Nr. 6, 1911, p. 1.

64. OBEL, M. DE L', Kruydboeck, Antwerpen, 1581.

65. PELLEW, C., Types of segregation. Journ. of Gen. 6, 1917, p. 317.

66. PUNNETT, R. C., On a case of patching in the flower colour of the sweet pea (*Lathyrus odoratus*). Journ. of Gen. 12, 1922, p. 255.

67. RASMUSSON, J., Genetically changed linkage values in *Pisum*. Hereditas 10, 1927—1928, p. 1.

68. RASMUSSON, J., Letalfaktorer hos ärter. Nordisk Jodbrugs-forskning. Ber. 4 Kongr., 1929, p. 611.

69. SCHERZ, W., Beiträge zur Genetik der Buntblätterigkeit. Zeit-schr. f. ind. Vererb. und Abst. 45, 1927, p. 1.

70. SIRKS, M. J., Multiple allelomorphs versus multiple factors. Proc. of the Intern. Congr. of Plant Sciences 1, 1929, p. 803.

71. SHULL, G. H., Über die Vererbung der Blattfarbe bei *Melandrium*. Ber. d. D. Bot. Ges. 31, 1913, p. (40).

72. SHULL, G. H., Duplicate genes for capsule-form in *Bursa bursa-pastoris*. Zeitschr. f. ind. Abst: und Vererb. 12, 1914, p. 97.

73. SÔ, M., On the inheritance of variegation in barley. Jap. Journ. of Gen. 1, 1921, p. 21.

74. SPRENGER, A. M., Witbontheid bij enkele groenten. Veldbode 14, 1916, p. 782.

75. STERN, C., Multiple Allelie, Handbuch der Vererbungswissen-schaft 1, G, Bornträger, Berlin, 1930.

76. STEHLÍK, V., Beitrag zum Studium der Abnormalitäten bei der Zuckerrübe. Zeitschr. f. Zuckerindustrie der Cechoslov. Republ. 45, 1921, p. 409.

77. STOMPS, TH. J., Über zwei Typen Weiszrandbunt bei *Oenothera biennis* L. Zeitschr. f. ind. Abst. und Vererb. 22, 1920, p. 261.

78. SUTTON, A. W., Compte rendu d'expériences de croisements faites entre le pois sauvage de Palestine et les pois de commerce dans le but de découvrir entre eux quelque trace d'identité spé-cifique. Compt. Rend. IV. Conférence Internationale de Gé-nétique, Paris, 1911, p. 358.

79. TAKEZAKI, Y., Lectures in crop breeding. (In Japanese). Tôkyô, 1922.

80. TAMMES, T., Das Verhalten fluktuierend variierender Merkmale

bei der Bastardierung. Rec. d. Trav. Bot. Néerl. 8, 1911, p. 201.

81. TEDIN, H., The inheritance of flower colour in *Pisum*. Hereditas 1, 1920, p. 68.

82. TEDIN, H. and O., Contributions to the genetics of *Pisum*. III. Internode length, stem thickness and place of the first flower. Hereditas 4, 1923, p. 351.

83. TEDIN, H., Eine mutmassliche Verlustmutation bei *Pisum*. Hereditas 4, 1923, p. 33.

84. TEDIN, H. and O. and WELLENSIEK, S. J., Note on the symbolisation of flower-colour factors in *Pisum*. Genetica 7, 1925, p. 533.

85. TEDIN, H. and O., Contributions to the genetics of *Pisum*. IV. Leaf axil colour and grey spotting on the leaves. Hereditas 7, 1925—1926, p. 102.

86. TEDIN, H. and O., Contributions to the genetics of *Pisum*. V. Seed coat color, linkage and free combination. Hereditas 11, 1928, p. 1.

87. TSCHERMAK, E. VON, Bastardierungsversuche an Levkojen, Erbsen und Bohnen mit Rücksicht auf die Faktorenlehre. Zeitschr. f. ind. Abst. und Vererb. 7, 1912, p. 81.

88. TSCHERMAK, E. VON, Examen de la théorie des facteurs par le recroissement méthodique des hybrides. Compt. Rend. IV. Conférence Internationale de Génétique, Paris, p. 91.

89. VILMORIN, L. Levêque de, Les panachures des fleurs, 1852, Société philomatique de Paris (Extraits des procès verbaux des séances). Paris, 1886.

90. VRIES, H. DE, Die Mutationstheorie. Leipzig, 1901—1903.

91. WELLENSIEK, S. J., Genetic monograph on *Pisum*. Bibliogr. Genetica 2, 1925, p. 343.

92. WELLENSIEK, S. J., *Pisum*-Crosses IV: The genetics of wax. Meded. Landb. Hoog. 32, 1928, p. 31.

93. WELLENSIEK, S. J. and KEYSER, J. S., *Pisum* crosses. V. Inherited abortion and its linkage-relations. Genetica 11, 1929, p. 329.

94. WHELDALE, M., Further observations upon the inheritance of flower colour in *Antirrhinum majus*. Reports to the evolution committee of the Royal Society. Report 5, 1909, p. 1.

95. WHITE, O. E., Studies of inheritance in *Pisum* II: The present state of knowledge of heredity and variation in peas. Proc. Amer. Phil. Soc. 56, 1917, p. 487,

96. WHITE, O. E., Inheritance-studies in *Pisum* IV: Interrelation of the genetic factors of *Pisum*. Journ. of Agric. Res. 11, 1917, p. 167.

97. WINGE, Ö., The chromosomes. Their numbers and general importance. Comptes-Rendus d. Trav. d. Lab. de Carlsberg 13, 1917, p. 131.

98. WINGE, Ö., On the non-mendelian inheritance in variegated plants. Comp. Rend. d. Trav. d. Lab. Carlsberg 14, 1919, No. 3, p. 1.

99. YASUI, K., Studies on the maternal inheritance of plastid characters in *Hosta japonica* ASCH. & GRAEBN. f. *albomarginata* MAK. and its derivates. Cytologia 1, 1929, p. 192.

EXPLANATION OF THE COLOURED PLATES

PLATE I.

Top to the left	1 : deep apple rose
Top to the right	2 : apple blossom
Centre to the left	3 : violet
Centre to the right	4 : rose
Bottom to the left	5 : violet leaf axil
Bottom to the right	6 : rose leaf axil

PLATE II

Top to the left	7 : apple dotted
Top to the right	8 : rose dotted
Centre to the left	9 : purple patched
Centre to the right	10 : F_1 purple dotted ($A_2\,a$)
Bottom to the left	11 : crypto purple
Bottom to the right	12 : purple dotted ($A_2\,A_2$)

PLATE I

R.Hooksema.del.

PLATE II

R.Hoeksema.del